Animal
Wonder World

A CHRONICLE OF THE UNUSUAL
IN NATURE

Animal Wonder World

A CHRONICLE OF THE UNUSUAL

IN NATURE

By Frank W. Lane

M. EVANS
Lanham • New York • Boulder • Toronto • Plymouth, UK

M. Evans
An imprint of The Rowman & Littlefield Publishing Group, Inc.
4501 Forbes Boulevard, Suite 200, Lanham, Maryland 20706
http://www.rlpgtrade.com

10 Thornbury Road, Plymouth PL6 7PP, United Kingdom

Distributed by National Book Network

Library of Congress Cataloging-in-Publication Data Available

ISBN 13: 978-1-59077-444-1 pbk: alk. paper)

♾™ The paper used in this publication meets the minimum requirements of American National Standard for Information Sciences—Permanence of Paper for Printed Library Materials, ANSI/NISO Z39.48-1992.

Printed in the United States of America

Preface

It would have been impossible to write this book without recourse to the experiences and observations of men on animals throughout the world. The extent of my obligation to the work of others may be judged from the 600 odd entries in the bibliography. To the scientists, field naturalists, sportsmen and others whose names appear therein I am deeply indebted.

The subject matter of several chapters is drawn largely from chance observations of animals by various witnesses, many of whom are not scientists. The source of nearly every observation is listed in the bibliography, so that readers can judge for themselves the authenticity of any given incident. The anecdotal method of nature writing has obvious drawbacks and limitations, but I think it is worth while to publish this material, if only in the hope that it will stimulate others to delve more deeply into some of the subjects dealt with.

Many of the works I have consulted are not easily available, and without the resources of great libraries this book could not have been written. I wish particularly to express my appreciation of the help I have received from the Library of the Zoological Society of London, and its capable and obliging Librarian, Mr. G. B. Stratton.

Believing in the principle embodied in the Chinese saying that "One picture is worth 10,000 words," I have included numerous illustrations. I should like to pay tribute to the patience and skill of the photographers whose work is here represented.

I have been considerably helped by the advice and criticisms of many scientists who have been good enough to read over various parts of the book. I desire to thank the following authorities who have kindly helped in this way:

The Rev. Edward A. Armstrong (Chapter 6); Dr. Geoffrey Beven (all the chapters); Mr. Eric A. Brook (Chapter 3); Dr. Frank A. Brown, junior (Chapter 2); Dr. H. O. Bull (Chapter 2); Dr. Colin G. Butler (the bee sections of Chapters 5, 6, 7, and 11); Dr. F. J. J. Buytendijk (Chapters 1, 7 and 11); Mr. Victor H. Cahalane (Chapter 13); Dr. Frances N. Clark (the grunion section of Chapter 7); Mr. Douglas Dewar (Chapters 4 and 5); Mr. James Fisher (Chapters 1, 6 and 8); Sir Hugh Gladstone (Chapters 4, 5 and 10); Dr. E. W. Gudger (Chapter 5); Mr. Martin A. C. Hinton, F.R.S. (Chapter 9); Dr. Carl L. Hubbs (Chapter 7); Mr. H. Roy Ivor (Chapter 8); Mr. David Lack (Chapter 3); Professor David Katz (Chapters 1, 7 and 11); Dr. W. L. McAtee (Chapter 10); Mr. Norman J. McDonald (Chapters 1 and 12); Mrs. Margaret H. Mitchell (Chapter 12); Dr. Robert Cushman Murphy (Chapter 10); the late Mr. J. R. Norman (Chapter 2); Dr. T. C. Schneirla (Chapters 6 and 11); Dr. Alan E. Slater (Chapter 1); Dr. W. H. Thorpe (Chapter 11); Dr. Ethelwynn Trewavas (Chapter 2).

I wish also to thank Mr. D. E. Brown, Mr. Charles and Miss Valerie Jiles, Mr. R. F. Newton, Mr. Frank Whitaker, and Miss Irene Thornley for advice and general criticisms.

In a book covering such a wide field, and dealing with so many different animals, it is almost impossible to achieve complete accuracy, but I have taken what precautions I could to eliminate error. It must be stated that ultimate responsibility for everything in this book is, of course, mine. I have not accepted every suggestion and criticism, and some material has been added after the critical reading.

I desire to thank the Editors of the following journals for permission to reprint all or part of my original articles: *A.T.C. Gazette, The Aeromodeller, Airman, Air Trails Pictorial, Animal and Zoo Magazine, Boy's Own Paper, Britannia and Eve, Country Life, Everybody's Weekly, The Field, The Listener* (text of broadcast talks), *Men Only, National Review, Nature Magazine, Natural History, Outdoors Magazine*, Pie Publications, *Toronto Star Weekly, The Strand, Tit-Bits* and *Zoo Life.*

Contents

Animal
Wonder World

A CHRONICLE OF THE UNUSUAL
IN NATURE

1

The Split Second in Nature

IF THE WRITERS of the cowboy sagas are to be believed, the action known as "the draw" is one of the fastest of all human movements. And scientific tests have certainly proved that an expert gunman can pack action into the unforgiving second.

Experiments have been carried out both by electrical stop-watch, and also with a stop-watch connected to an electrical device which noted the time the hand touched the butt of the revolver and the time the bullet was fired. Chauncey Thomas, one of the greatest of the old American frontier revolver experts, once pulled a belt gun and hit a target ten yards away in three-fifths of a second. This was an exceptional performance. The average time, proved by hundreds of trials, required by a man to draw and shoot was one and two-thirds seconds. (Roth.) During some police tests in this country a G-man drew his revolver and put one shot into each of three targets in one and three-tenth seconds. (*Popular Science Monthly,* May, 1940.)

In more recent tests Delf Bryce, of the F.B.I., dropped a dollar at forehead level and then, with the same hand, drew his revolver and was ready to fire as the coin reached his waist. Stroboscopic photographs illustrating this feat appeared in *Life* for November 12, 1945.

Fast as these actions are, there are other human movements which rival them. The human eye has been proved by high-speed cinematography to blink in one-fortieth of a second. During the "mad minute," when a first-class typist types and re-types a simple sentence for sixty seconds, the keys and spacebar are sometimes tapped sixteen times per second. A human sneeze rarely lasts more than a tenth of a second, and some of the droplets in a hearty sneeze are shot out at a speed of one hundred and fifty-two feet per second, or just over one hundred miles an hour! (Jennison.)

Timing by photo-electric cell has proved that a boxer's fist can flash out with such acceleration that in the last few inches of travel it can reach forty miles an hour. (*Popular Science Monthly*, September, 1938.) This helps to explain why boxers have sometimes received knock-out punches which travelled less than a foot. Apropos this speed of the human hand, Frances Pitt writes: "My brother was standing in a gateway, when he saw a sparrow-hawk flying down the hedgeside towards him, when he instinctively and without thinking threw up his hand as if to catch a ball, and caught the bird instead! He was too startled to hold it, and it dashed off again in a great flurry and hurry. He said he did not know which was the more astonished, he or the bird!"

The relation between such fast actions and the mental operations accompanying them make an interesting study. The mind "moves" at lightning speed compared with the body. Mountaineers who have been involved in accidents have recorded that during the split seconds between their realisation that a crash was impending and the crash itself, the mind has leisurely surveyed various incidents of the past, tried to recall things half-forgotten and weighed up the possibilities of the coming crash. (Amery.)

Similarly it is recorded that when Battling Munro was knocked out by Sam Langford, an entire novel unfolded itself to his subconscious mind during the ten seconds of the count! (*The World Says*, September, 1938.) It may be remembered that Ambrose Bierce's short story of the Civil War, *An Occurrence at Owl Creek Bridge*, is based upon what can flash through a man's mind in a fraction of a second. In this story a Federal spy is hanged from a bridge. While he is dropping to his death he imagines that the rope breaks, he escapes by swimming under water and makes his way back to his own lines.

An appreciable time elapses between the mind's command to "go" and the consequent action by the muscles. This elapsed time is generally called reaction time. If an experience of my own is any guide, reaction time can, on occasion, be realised as a definite sensation. A friend once tested me on a home-made reaction tester in which a switch had to be pulled when a light signal was given. The time elapsing between the switching on of the light and the pulling of the switch was recorded by the tester.

On one occasion my mind had wandered slightly and when the light glowed I was unprepared. I remember my mind commanding "Pull the switch!" and then I realised a feeling of impotence until the message had reached my hand. I wanted to pull it, but had to wait until the command got through.

Tests have been made on the times needed by automobile drivers to see a danger signal, realise its meaning and commence to apply the brake. In one series of tests, 2,245 observations were made. The average reaction time was 0.62 of a second. Five per cent. of those tested took one second or even longer. These figures were obtained under road conditions.

In tests which have been made under laboratory conditions, better figures have been obtained. A driver who took as long as one and a half seconds to apply the brakes on the road took only 0.2 or 0.3 of a second in the laboratory. These figures are especially interesting in view of the general belief that a driver reacts more quickly in a real emergency than in a formal test. (Lester.)

What these reaction times mean in terms of traffic safety (or danger!) may be gauged from the fact that a car travelling at sixty miles an hour covers 44 feet in half a second, or less than the time most drivers require to realise an emergency and begin to apply the brake.

Tests have also been carried out on trigger reaction time. They were made with the aid of the new radio-tube super stop-watch known as the chronoscope. These tests revealed that if a rifleman has his rifle aimed and is watching for a target to appear a hundred feet away, it will take about a quarter of a second before the bullet strikes it.

This time has been analysed and found to be made up of the following time-intervals. A hundred and sixty milliseconds are taken up by the man between the time the target appears and the pulling of the trigger; for the trigger and firing mechanism to operate ten more milliseconds are required; two milliseconds are taken up while the bullet speeds down the barrel, and forty milliseconds later the bullet has travelled the hundred feet to the target, making a total of two hundred and twelve milliseconds, or just under a quarter of a second. Two-thirds of this time is taken up by the reaction time of the man behind the trigger. (Caldwell.)

Highly trained athletes probably have the quickest reaction times. When Babe Ruth, the famous baseball player, was tested at the psychological laboratories of Columbia University it was found that his reaction time was half that of the average. Such speed in reacting was of great advantage to him when facing a fast pitcher, as it enabled him to wait and study the ball longer before starting his swing. And in baseball you can lose or win in one-hundredth of a second! (Burpee and Stroll.)

Burpee and Stroll have carried out a large number of reaction time tests with athletes. They were made with both small- (the hand) and large-muscles (arms, legs, etc.). The very fastest small-muscle time was 0.1005 of a second and the fastest large-muscle time was 0.56 of a second. In the large-muscle tests, however, the subject had to move six feet and reach forward his hand. These times are a confirmation of the assertion that first-class sprinters, although appearing to the human eye to take off at the crack of the starter's pistol, actually take a tenth of a second to start moving. And at the end of a hundred yards race a first-class sprinter can cover three feet in that one-tenth of a second.

The average reaction times are: Sight, 0.19 to 0.22 of a second. Hearing, 0.12 to 0.18 of a second. Electrical stimulation of the skin, 0.12 to 0.20 of a second. (Evans.) In Seashore and Seashore's series of experiments with male subjects a faster average reaction time (0.138 of a second) was achieved with the jaws (biting on a key when a sound was made) than with either foot or hand. But this faster time with the jaws did not obtain with female subjects.

The importance of reaction time was many times demonstrated during the Second World War, notably in air combat. An instance of it was furnished in a Stuka *versus* destroyer action. The captain of the destroyer had himself strapped to his bridge and kept his glasses fastened on the attacking planes. The moment he saw the bomb leave the Stuka he ordered the destroyer to change course. If everybody concerned reacted quickly enough the destroyer could be a hundred or more yards away by the time the bomb fell.

High-speed films have revealed how in physical, as opposed to anticipated emergencies, reaction time is also of an appreciable duration. I have seen a film which shows that when a strong light is flashed on to the eye the instinctive blink comes *after* the light has been extinguished. In the same film a burning cigarette is seen being applied to a man's hand. The cigarette is pressed into the hand and completely withdrawn before the hand begins to move.

It is rarely possible to subject wild animals to the exact tests which can be applied to human beings, but a number of observations have been put on record by scientists and other watchers of animals which provide interesting comparisons with what man can do in the realms of fast movement and reaction time.

A film was once taken at the London Zoo, of a toad swallowing a worm. A high-speed camera was used, with the mechanism adjusted to take three hundred pictures a second. When all was ready, and the camera was running, a worm was placed in front of the toad. The next thing the cameraman knew was that the worm had vanished.

The whole action—the tongue flying out, gathering the worm and returning to the mouth—was said to take place in a small fraction of a second. Just how small a fraction may be gathered from the statement by the experienced student of toad life, Jean Rostand, who says: "The speed of lingual projection is considerable. The double journey, there and back, takes less than a fifteenth of a second."

Another animal renowned for the extreme rapidity of its tongue movements is the chameleon. It has been described as possessing "the most amazing tongue in Nature," and when the tongue's nervous and muscular organization and its mode of working are studied it is realised that such a description is well merited. (See Murphy, Zoond, and Gnanamuthu for details.)

Murphy says: "The plan of the weapon might be said to combine certain features of the stretched rubber bands of a slingshot with the extraordinary propelling force obtained by squeezing a wet watermelon seed from the tips of thumb and forefinger!" He adds that a chameleon with a seven-inch body can shoot a fly twelve inches from its nose. For such an action the time required is in the region of half a second.

The strike of some snakes is also very rapid. It has been asserted that a cobra attacks with the speed of a "cracking whip" and that the strike of a rattlesnake "is the fastest thing in Nature." Both statements are probably exaggerations, but the famous herpetologist, Dr. Raymond L. Ditmars, says: "Stroke, bite, pressure on the poison glands, injection of the venom and return of the head to the snake's defensive position normally takes less than half a second." Considering the movements and actions involved this is certainly very rapid action.

But if snakes can move at this speed, what shall be said of the movements of some of their conquerors, such as mongooses and some of the lizards, which can not only avoid the snake's vicious lunges, but can get in their own bites without being bitten in return? Dr. F. J. J. Buytendijk, who has made a special study of cobra and mongoose fights, informs me that the two animals move together just like two hands that collaborate. There is no reaction time between one movement and the other, for all that occurs is anticipated by the other animal.

Teale, who has watched the strikes of various snakes in the New York Zoo, says the strike of a cobra is relatively slow. He writes: "It wasn't in the same class with the low, upward drive of the rattler. It was easy to see how the mongoose, effective as a killer of cobras in the Old World, was helpless in the face of the lightning drive of the venomous snakes of the New World." (But a mongoose has been known to kill a fer-de-lance—see *Natural History*, February, 1947.) Smith points out also that cobras see very badly during the day-time.

The gaff of a fighting-cock is another extremely rapid action. It is claimed that no human eye is quick enough to record that lightning blow strike home, and no ordinary camera can "freeze" it. Some indication of the speed of the gaff can be gathered from the fact that a cock has been known to drive its steel-mounted spur half an inch into a well-seasoned oak board.

In addition to moving parts of their bodies at high speed, many animals are also capable of hurling themselves into action in a very short space of time. Such an ability is not surprising when it is remembered that in wild Nature a split second is often all that divides the quick and the dead. In fact, Marcus Daly, a professional big-game hunter with great experience of the carnivora, says a game-eating lion will kill a buffalo, zebra, eland or other game in less than half a second.

As far as I am aware, the fastest movement found in any animal is the discharge of the electric eel. According to Rosenblith and Cox, an electric impulse flashes from tail to head of the eel at a speed of half a mile a second, or possibly even faster.

A famous correspondence took place in the columns of *The Times* (London), in September and October, 1931, on the acceleration attained by several large animals. The well-known big-game photographer, Marcuswell Maxwell, wrote that he had been charged by a gorilla which had leapt from a tree. Maxwell estimated that the animal had covered fifteen yards in under a second (and any experienced photographer, such as Maxwell, knows that a second is a long time).

A subsequent writer in the correspondence challenged such an estimate. In his reply, giving his reasons for believing that his figures were accurate, Maxwell said that:

> . . . having seen how the greater portion of the gorilla's weight is contained in his enormous chest, shoulders and arms, I am not now surprised at it [i.e. the gorilla's acceleration]. Nor, having seen how his arms and legs cut up the ground where he started out, ground which, although moist, was solid and matted with innumerable tree roots, do I feel inclined to state a limit to the acceleration which he may attain at the beginning of such a rush, always given suitable trees or other objects on which he can get a purchase.

During the course of the correspondence one person wrote: "No living creature could cover fifteen yards in well under the second." This statement brought a reply from a man who had spent years in Africa and said: "From my own experience I should say that a leopard could cover fifteen yards in a second, springing from a standing start. They move literally like lightning." (No, not "literally," i.e. some 28,000 miles a *second!*)

Confirmation of the extreme agility of the leopard comes from Pohl, who relates how he watched a leopard creep towards a duiker antelope and then make its attack. Pohl says: "So incredibly swift was his act that fifteen yards must have been covered in about half a second; but the slight noise he inevitably made in doing so roused the duiker to instant action, for he leapt yards away from where he had been standing the minutest fraction of a second before."

Sutherland writes of the leopard:

> No animal can surpass the leopard in concentrated fury of attack when thoroughly aroused, and when in this state it behoves the hunter to exercise extreme caution. No one who has not experienced it can have any notion of the speed with which they charge. The uninitiated might fondly imagine that a chance of dodging might be possible, but such is not the case. A streak of light cannot give much start to a leopard's charge.

Super-gymnast though he is, even the leopard must give way to the cheetah for sheer speed from a standing start. The acceleration of a prime cheetah is said to be forty-five miles an hour in two seconds from a standing start. (*The Times* (London), June 30, 1939.) And when in its full stride it can cover a hundred and three feet in a second (seventy miles an hour).

It is of interest to compare such speeds with those of a first-class human sprinter. In two seconds he can reach a speed of only sixteen miles an hour, and when going all-out (i.e. in the middle of the hundred yards sprint) he can reach only twenty-four or twenty-five miles an hour.

It is not only the larger and more spectacular mammals which possess such remarkable powers of acceleration. Lizards are experts in flashing into action and also in coming to a dead-stop after a lightning burst. Alfred Mosley tells of coming upon a monitor fast asleep in the Kalahari Desert. Mosely came to within a yard of it, then clapped his hands. Then "there was no sign of the animal, but simply a streak of dust right away to the horizon" (in a letter from the late Captain Guy Dollman).

Another lizard renowned for its great powers of acceleration is the six-lined race-runner (*Cnemidophorus sexlineatus*). Pope thus writes of its sprinting powers:

> The chief characteristic of this lizard is the astonishing speed of its dashes for safety; at such times it appears as a streak across the ground, and indeed is called "streak-field" in Georgia. If these dashes do not end at its burrow, the lizard will halt in cover as suddenly as it began and thus seem to disappear. . . . The speed of the wild dashes has not been accurately determined, although it is claimed that a rate of eighteen miles per hour may be attained.

And eighteen miles an hour for a creature which, irrespective of its tail, rarely measures more than three inches, is certainly fast moving.

As may be expected, in encounters between birds of prey and their would-be victims the role of the split second is frequently illustrated. Savage records an incident which he observed when a quail was flying for its life before a pursuing Cooper's hawk. The hawk was gaining on its quarry when the quail, from a height of about six feet, dropped like a stone towards a clump of hazel bushes, which offered it sanctuary from its foe. But the hawk, hurtling itself through the air, flung its body beneath the quail, turned upside-down and spread wide its talons to receive the quail's body as it fell. Righting itself, the hawk flew away with its prize.

On another occasion a black pigeon hawk, by dashing tactics, neatly robbed a sportsman of his bag. He shot a snipe, which started to fall, apparently with a broken wing. Before it reached the ground, "from nowhere" there flashed a hawk which travelled so fast that it could be seen only as a "blurr in the air." The hawk caught the snipe in mid-air and flew away so swiftly that it was almost out of gun

range by the time the surprised sportsman realised what
had happened.

Such an incident is by no means isolated where these
dashing buccaneers of the air are concerned. Another
sportsman shot a plover and his dog went to collect it. But
suddenly, before even the report of the gun had died away,
there dashed through the smoke a sharp-shinned hawk
and picked up the plover in front of the dog's nose. Again
the hawk got clean away before the man had recovered his
wits sufficiently to fire a forlorn shot at the robber. (Bent,
1937 and 1938.)

The perfect ease with which birds of prey sometimes
capture their quarry illustrates, not only extremely rapid
movement in Nature, but also wonderful co-ordination of
eye, wing and talon. A sparrow hawk has been seen to
swoop over a pond and pick off a swimming moorhen
without even making a splash. An Eastern sparrow hawk
in rapid flight was seen to pick off a "lizard" from the
trunk of a tree without pausing in its flight. A peregrine
falcon will sometimes gather up a small bird so neatly and
swiftly that no alteration in its flight and no act of seizure
can be detected.

But it is not only in the taking of life that birds of prey
execute the lightning manoeuvre. Often their dashing
tactics lead them into situations where only the quickest
instinctive action can save them from serious injury or
death.

Gill describes a headlong chase he witnessed between a
peregrine and a starling. The starling managed to save
itself in the nick of time by diving into a bush. "The peregrine saved itself by an impossible-looking turn which
shot it into the sky like a rocket."

A correspondent (F. H. Pearce) has told me how he

once watched a sparrow hawk perform a perfect Immelmann turn. The hawk was flying at full speed after a finch, which was already crying out in the black minute before capture, but it managed to escape by diving under a large wall of telegraph wires and found sanctuary in some scrub. The hawk, with no eyes for anything save its victim, suddenly found itself hurtling into the bank of wires, which would have cut it to pieces at the rate at which it was travelling. To save itself, the hawk shot straight up into the air like a rocket, turned into a half-loop, rolled over at the top of the turn and flew back the way it had come!

I am not certain whether the following action of a sparrow hawk should be interpreted in the same way as the incident cited above, namely, as an intelligent, if largely instinctive, effort to save itself from disaster by the execution of a lightning manoeuvre, but the incident is too remarkable to be omitted here. It was told to me by an eye-witness, Noel M. Sedgwick, the editor of *The Shooting Times.*

A sparrow hawk was flying low down a ride in a wood. At the end of the ride was a five-barred gate. Waiting by the side of it was a keeper with his gun trained just above the top of the gate. His intention was to shoot the hawk as it flew over the gate. But as the hawk reached the gate, instead of flying over, it flew *between* the fourth and top bars! Did the hawk, as it came to the gate, catch a lightning glimpse of the waiting keeper, and with a speed of reaction and muscular co-ordination that baffles understanding, take the only course which could save its life?

Priest tells of an incident he witnessed which exemplifies another aspect of the split second as applied to birds, namely, the ability to stop dead in a fraction of a second after travelling at high-speed. He saw a Wahlberg's eagle stooping from a great height on to a black-breasted harrier-eagle which was carrying a snake. When only twenty feet above the harrier-eagle, the Wahlberg braked suddenly with wings and tail.

The effect, says Priest, was magical. "From a speed of anything up to a hundred miles per hour she came to a dead stop in that small area. Her brakes were so accurately applied that she had momentarily ceased to move when parallel with the feet of the harrier. In a flash his breakfast had gone! This little robbery took place about two hundred feet up."

Despite their extreme agility, it is not birds, but insects, who are the supreme masters of manoeuvrability in the air. The late Professor A. Magnan, of the Collège de France, studied insects in flight by the aid of ultra-high-speed cinematography. He obtained film records taken at nearly thirty thousand pictures per second (not per minute!). When the films were developed the images were drawn on paper on a larger scale. It was then possible to analyse aerial movements which it would have been impossible to appreciate by any other means.

It was thus that Magnan discovered that in one film sequence he had caught a hover-fly performing a somersault in mid-air in one-hundredth of a second. Magnan found other aerial manoeuvres had been carried out in one-thousandth of a second.

Burr gives an example of very rapid movement on the part of an insect not generally regarded as remarkable for speed of movement—an earwig. He says he put a bluebottle into a small glass-topped box containing a large earwig (*Labidura riparia*). Immediately the buzzing of the bluebottle ceased. Burr could detect no movement, yet in a flash the earwig's long forceps had transfixed the fly.

Some members of the Arachnida can also move with extreme rapidity at times. North tells of seeing a companion strike with his spade at a "tarantula" (probably one of the large hairy spiders which are popularly known as tarantulas in Texas). He narrowly missed, and the tarantula shot up the blade and handle of the spade like a streak. It was only by instantly dropping the spade that the man avoided the tarantula running on to his hand and from thence probably up his arm and on to his face.

Turning from these examples of rapidity of movement to reaction time among wild animals, it is of interest to note that, in some species, the time-lag between decision and action is apparently much less than it is with human beings. The ability to dodge shot-gun pellets and rifle bullets fired at fairly close range must surely be the severest tests of a creature's reaction time that could possibly be devised. Yet some mammals and birds can pass even these ultimate tests.

In the old flint-lock guns there was an appreciable delay between the pulling of the trigger and the departure of the shot from the barrel. Experiments have proved that this delay varied from 0.075 to 0.105 of a second, with an average delay of 0.094 of second. With modern shot-guns the delay is very much less, about 0.004 of a second, or less than one-twentieth of the delay in the flint gun.

A charge of shot from a modern gun takes about one-seventh of a second to travel forty yards. A modern rifle bullet covers the same distance in about one-third of that time, or one-twentieth of a second. (Gladstone, and Caldwell.) These measured split seconds should be borne in mind when considering the following examples of the reaction times of various animals.

Old Charles St. John, writing of shooting at seals, says:

So quick are their movements in the water that I find it impossible to strike a seal with ball if he is watching me, for quick and certain as is a detonating gun, they are still quicker, and dive before the ball can reach them. As for a flint gun, it has not a chance with them.

According to Manly Hardy, the Canadian authority on the otter, these animals can similarly dodge a rifle bullet. Hardy says:

There are few animals as quick as an otter. When looking at a person, they will dodge a bullet as quickly as a seal can. I once fired at one which had his head out of his fishing hole in the ice. The ball struck not six inches beyond in the exact range of the centre of the hole, but the otter had drawn his head down before it struck.

And, writing of the weasel, Seton says it can "elude [a] rifle ball at the flash."

I discussed this subject with a man who had done a fair amount of shooting in Africa and he thought what I said threw a light on something which had often puzzled him. He said that although he had often shot at crocodiles resting on the water, he had rarely hit one. Up to the time of our conversation he had always thought it was his bad shooting which was responsible. But in view of what I told him of other animals dodging bullets, he thought perhaps the crocodiles had the same instant reaction to the flash of a rifle and had pulled their heads underwater by the time the bullet reached them.

This ability to dodge shots is possessed also by several species of water birds. Fire a gun at a cormorant resting on the water and, although the bird is within thirty or forty yards when the trigger is pulled, such is the speed of its crash-dive at the flash of the cartridge it is below the surface by the time the pellets reach it. (Perry.)

Philip Gosse, the famous Victorian naturalist, writes as follows of the Mexican grebe:

> On the slightest alarm they dive with the quickness of thought, and so vigilant is their eye and so rapid their motion that, ordinarily, the fowling piece is discharged at them in vain. It is commonly said of some birds that they dive at the flash of the pan; but though I always used percussion locks I could never succeed in hitting one until I formed a screen of bushes, behind which I might fire in concealment. I then found no difficulty. Hence, I infer that their quick eye detects and takes alarm at the small but sudden motion of the falling hammer.

A similar ability to dodge at the flash of the gun is related by competent authorities of the bufflehead, oldsquaw, Atlantic harlequin duck, murre, Mandt's guillemot and Holboell's grebe. (See Bent, 1919 and 1925, and Kortright.)

But the ability to dodge bullets is found in the highest degree in the loon. Dawson tells how he once concealed himself and shot seven times with a Winchester repeating rifle (not a shotgun) at a loon on a pond. The bird beat him every time by its crash-dives, although, of course, some of Dawson's shots may have been wide.

Dawson says: "Generations of gun practice have made the bird such an expert diver that, given room enough in which to dive, it is all but impossible to shoot one."

That veteran and widely experienced ornithologist, Arthur Cleveland Bent (1919), tells an even more remarkable, though pathetic, story of a concerted attempt to outwit a loon. He writes:

> I once saw a remarkable exhibition of this power of dodging at the flash of a gun by a loon which was surrounded by gunners in a small cove on the Taunton River. There were six or eight men, armed with breech-loading guns, on both sides of the cove and on a railroad bridge across it, all within short range. I should not dare

to say for how long a time the loon succeeded in dodging their well-directed shots, or how many cartridges were wasted before the poor bird succumbed from sheer exhaustion; but it was an almost unbelievable record.

I sincerely hope that no words of mine will induce any sportsmen to test further these remarkable powers of the loon, whose numbers have already been sadly depleted.

2

Fish with an Ear for Music

INTELLIGENT action and reflex action are controlled by different centers of the brain. In fish the center responsible for conscious action is relatively small. In reading these various accounts of experiments and experiences with fish, it should be borne in mind that not all actions which, to man, appear to be intelligent are necessarily so, but may be the result of conditioning built up over a period of time and resulting in a conditioned reflex to given stimuli. Nevertheless, he would be a hardy interpreter of animal behavior who would deny all vestiges of intelligent thought in everything that follows concerning the actions and reactions of fish.

The somewhat bizarre title of Professor J. P. Frolov's book, *Fish Who Answer the Telephone,* was but an imaginative description of happenings which the author had witnessed in his own laboratories. Frolov tethered fish in a small aquarium upon a light, flexible electric wire, with plenty of slack to enable the fish to swim freely. When a key was depressed a telephone receiver, submerged in the aquarium, emitted a sound and a slight electrical shock was transmitted to the fish through the tethering wire. The fish reacted to the shock by making an agitated movement in the water.

After some forty trials Frolov omitted the shock but sounded the telephone signal. The fish reacted with the agitated movement as in the previous experiment. They had learnt to answer the telephone! Even when a bell was rung above the water, the fish responded in the same way. In trials carried out by other experimenters it was learned that fish could be trained to move into an under-water dining-room when their dinner-gong was sounded.

These experiments on the hearing and intelligence of fish have been carried a stage further by von Frisch, and Stetter. They placed several blind minnows in an aquarium and then trained them to associate a given sound with food. These sounds were given by whistles and tuning-forks.

After the minnows had learned the meaning of the feeding-sound they invariably reacted to it; some by diving down and snapping for food on the bottom, others by swimming to the surface and snapping there, and others by stopping and snapping vigorously.

The investigators then endeavored to find the acuteness of the minnows' hearing. It was proved that a soft sound, given 200 feet from the aquarium, produced the feeding-reaction. Then a large aquarium was placed alongside that in which the minnows were kept and a man dived down inside. The sound was given again and, as judged by their reactions, some of the minnows heard as well as, or even a little better than, the man in the larger aquarium.

The next step was to see if the minnows could distinguish between sounds and, if so, to what extent. Again a minnow was taken and thoroughly trained to the feeding-signal. Then another sound was introduced, one or two octaves higher or lower than the feeding-signal, but no food was given. If the fish reacted to this signal with the feeding-reaction it was punished by a light blow with a glass rod. It was found after some time that the minnow gave the feeding-reaction only to the feeding-sound, and when the other signal was given, it either gave no reaction or else a distinct flight-reaction to avoid the expected punishing blow.

Most of the fish used in these experiments could easily be trained to distinguish between two sounds at an interval of one octave, and some fish could distinguish between two sounds a minor third apart. Other experiments showed that some of the fish could differentiate between notes and noises, two alternating tones, and between four to five different tones. But differentiation between varying intensities of the same tone could not be proved.

A lady, with whom I was discussing these experiments, told me of the remarkable behaviour of two carp which she kept in a small aquarium in her living-room. She said she had repeatedly noticed that when dance music was being played on the radio the carp "danced" up and down, but that when music without any dance rhythm was played there was no such response on the part of the fish.

Whatever may be thought of this report, it was proved in a series of experiments by Parker, that when a bass viol string was vibrated near an aquarium containing minnows, the fish responded by rapidly vibrating their pectoral fins.

Bull, who has made a detailed investigation on conditioned responses in fishes, found that he could train them to respond to changes in the water of the aquarium. A blenny learnt to rise to a feeding-place in response to a momentary increase of four degrees centigrade in the temperature of the water, or to an increase of 3-1000 in the salt content of the water. When these changes were introduced after the fish had been trained, it swam to the feeding-place even when no food was offered.

Herter trained a catfish in touch stimulation. The fish learned to distinguish between a rough and a smooth glass rod when either of these touched the margin of its mouth.

Several series of experiments have been carried out in recent years on vision in fish. Some of the results achieved were no less startling than those of the experiments on hearing.

Dr. Frank A. Brown, of the Illinois Natural History Survey, made over 14,000 trials with large-mouth black bass. A small glass tube, with a coloured band wrapped round it, was lowered among the fish. When a fish swam near it food was given. Then other colors were wrapped round the tube, but if one of the bass came near these a tiny shock was given by touching the fish's back with a wire powered with a mild electrical charge.

It was found that after five to ten trials most of the fish could distinguish between red, yellow, green and blue in strong pure colors. The experiments were then carried a stage further. Shades of the colors were wrapped round the signalling tubes and the experiments were continued until the limit of the fishes' vision in color differentiation was reached.

Brown concluded: "It is probable that large-mouth black bass are able to distinguish among colors in about the same manner as would a human being with perfectly normal color vision, looking through a yellowish filter."

In addition to the facts concerning the vision of the bass which these experiments proved, they also provided an indication of the length of these fishes' memories. Once the fish had learnt to distinguish the "right" and "wrong" colors, this knowledge persisted for several weeks and with some fish for over a month.

Experiments with other fish have shown that they, too, can be taught to distinguish between different colors. Gertrude M. White placed food at the end of a pair of forceps, with a red cardboard disc fixed immediately behind, and then held them just above the surface of some water containing minnows and sticklebacks. After a while the fish jumped out of the water and seized the food.

Some pieces of paper were then substituted for the food and a blue disc replaced the red one. At first the fish jumped for this just as they had for the red disc with the real food. But after a while some of them learned their lesson—they made no attempt to jump for a blue disc, but leaped with open mouths when the red disc was shown.

In another series of experiments with some mud-minnows, traffic lights were used to give the signals. The fish were trained to jump whenever a red light was flashed; when green was shown they were trained to retire to one of the bottom corners of the aquarium. Within two months a shoal of fish, trained one by one, had learnt their lesson perfectly. But when an attempt was made to train several fish in the same aquarium at the same time, the rate of learning was slower. Perhaps, like some human beings, fish learn more quickly by private tuition! (Allee.)

Wolff's experiments with minnows, in which he used spectral colors of controlled intensities, proved that some of these fish could distinguish between twenty color tones.

These and other experiments go far to prove that many species of fish are able to distinguish between various colors. The experiments of Herter have shown that, in addition to this color vision, fish have a good eye for form. Herter had a "class" of twenty-one fishes, composed of six species. He trained them to respond to various optical signals composed of differently shaped cardboard figures. When a fish made a correct response it was rewarded with food, but when a wrong response was made it was given wax.

The best results appear to have been obtained with minnows and perch.

> After training to circles as against ellipses, the circle being positive, the fish [minnow], when confronted with a circle and also an ellipse of different comparative axes, always chose the circle. When various ellipses were offered to him, he chose the most nearly circular. The fish trained to an ellipse as a positive signal chose in the test the ellipse to which he had been trained; when this was not presented he chose the signal which resembled it most closely. Thus the fishes made an absolute choice when this was possible. When the positive signal was lacking in the test they chose the one nearest to it in height and breadth. . . .
> Training to the letters R and L succeeded well [with perch]. The fishes learned to distinguish the letters according to their position and shape, and orientated themselves consistently.

The ability to distinguish even small differences in form is well illustrated by two experiences which Bower-Shore had with perch which he kept in one of his aquariums. When he approached one of the perch with the small worm-box from which it was fed in his hands, it showed considerable agitation. When, however, he entered the room without the box, there was no similar reaction.

Bower-Shore had another perch which had been taken with a net from a pool. It bore no trace of having been injured by a hook. Despite its apparent lack of experience of the deadliness of the hook, whenever a hooked worm was placed in its aquarium the perch would not take it. But when the worm was suspended by a thread only, the perch rushed to take it.

This ability on the part of a fish to recognise even slight differences in form, as indicated by Herter's experiments and Bower-Shore's experiences, renders more credible some of the stories which have been told in the past of fish recognising various people.

A lady used to keep goldfish and minnows in a small artificial pond. At feeding-time she put her hand in the pond and both goldfish and minnows swam over to her and jostled one another in their endeavour to take the food from her fingers. One day a visitor tried to imitate her, but, although she acted in exactly the same way, the fish ignored her. (Keay.)

The owner of the pond said she had had a similar experience when she had tried to obtain a response from the occupants of a fish hatchery. Again the fish failed to respond, although they did so every time to their attendant. Incidentally, a lobster, that used to be kept in a small tank in the Forest Hill Museum, in England, always put a claw out of the water whenever its keeper came along to feed it. (Martin.)

Another example of the ability of fish to distinguish between those who feed them and strangers is given by Clarke. He says a friend had a trout stream in his grounds, but he did not permit fishing and treated the trout as pets, feeding them daily. Clark writes: "They would follow him up-stream in a shoal. If I walked with him they took no notice, but if I went alone, even if armed with food, they all bolted into cover at once, thus clearly distinguishing between their friend and protector and a stranger."

One of the most successful fish-tamers of all time must be Raul Vasquez, of Key West, Florida. In a tidal pool, 280 feet long by 80 feet wide, he keeps an odd assortment of more than seventy varieties of fish.

"I do a lot for my fish," he says, "that is, I provide a nice comfortable home for them and never allow them to be annoyed—by myself or anyone else. I give them plenty of good fresh food and when I tap on the rock they understand: they know it is feeding-time. Within a week I tamed a fierce barracuda so that it became as tractable as a pet cat."

Vasquez talks to his fish in almost the same manner as others talk to their dogs and horses. When he whistles fish jump to snap at morsels of food he holds to them. When he goes down to the pool and raps with a stick, fish swim to him from all directions.

An "impossibility" which he has accomplished is to get fish to allow him to pick them up. A man who has seen him do this says the fish remain quiet. With a final scratch on the back Vasquez returns them to the water.

Such a fondness for fish reminds me of the report which appeared in a Newfoundland newspaper of a man who was so fond of his pet fish that it cost him his wife! According to the report, the wife applied for a divorce on the grounds that her husband insisted on keeping a large catfish in the only bath in the house.

There are a number of records of fish that have been trained to do simple tricks. Some of these stories are no doubt exaggerations of incidents which are susceptible of more simple explanations. But in the light thrown on fish behaviour by controlled experiments in zoological laboratories, there would appear to be nothing inherently improbable in the accounts which follow.

At the Roman baths in Bath, England some years ago, the authorities fixed a little platform, covered with ants' eggs, over a pool containing goldfish. A piece of string was attached to one end of the platform, and its free end dangled into the water. The goldfish soon learnt that, to operate this open-air cafeteria, all they had to do was to swim up to the string and give it a tug. This, of course, tilted the platform and a shower of ants' eggs fell into the water.

The owner of another goldfish trained it to swim to and fro through a small hoop which used to be lowered into the water. Incidentally, there was once a vaudeville act in which trained goldfish performed a few simple tricks.

During a B.B.C. broadcast on January 23, 1937, a Mr. McPherson who used to keep a large variety of fish for exhibition purposes, described how he taught one of them (he called it a higoi) to loop the loop in the water. Asked how he trained the fish to do this, he replied: "By trailing a worm through the water in front of him in a circular movement and making him chase it. Now, whenever he sees me, he loops the loop till he gets his worm."

The most highly trained fish I have read of was a carp which was an inmate of a fountain-basin in a park in California. This fish would respond to a call when its attendant whistled; swim forward, backward, or on its side through a hoop placed in the water; and even come out of the water into its master's open hands. (Berridge.)

In addition to the evidence concerning the intelligence of fish provided by these examples of training, a number of incidents concerning wild fish have been recorded which bear marks of intelligent behaviour.

Abbott says he was once standing near a small stretch of water when a gill-net was placed across the only outlet. In the enclosed waterway were several large pike. On being disturbed, six of these pike rushed towards the net.

The first pike struck the net and became securely entangled. Immediately the other five fish stopped dead and each of them appeared to solve the problem which suddenly confronted it in the way it thought best. One pike rose to the surface and, after pausing a moment, turned on one side and leaped over the cork-line. Another pike followed it.

A third pike swam close to the side where Abbott was standing and, discovering a narrow space between one of the ropes holding the net and the net itself, felt its way slowly through, although the water was so shallow that fully one-third of the pike's body was out of the water during this manoeuvre.

The two remaining fish turned back from the net and swam away. But being disturbed by a man splashing in the water, they again turned and swam towards the net. As their snouts touched the net they stopped. Then both fish suddenly sank to the bottom of the stream and burrowed into the sand, underneath the lead line, and in a moment reappeared on the other side of the net and were gone.

Abbott adds: "There was something in the manner of these fish, too, which is not readily described, but which gave an importance to those acts, on their parts, that I have mentioned, and which added materially to the strength of the evidence that they were 'thinking' in all that they did."

A more surprising example of "thought" in fish is provided by an experience of Holder's. He was fishing from a pier and hooked a fish which he calls a yellowtail. It ran out 200 feet of line before he stopped it. Then the fish turned and swam swiftly back towards the pier, dashed beneath it, secured a purchase on the line and broke it.

Now if the story had ended there one would be entitled to say that, although the fish adopted the only course which would give it its freedom, it had done so by chance. However, Holder continued fishing. He could see the fish that had broken loose swimming about below him. It was easily recognised because it was towing the broken end of the line and had a blotch on its jaw which had been made by the hook.

After another twenty minutes this fish again seized the bait. And this time, without making any preliminary run, it took a turn round one of the piles supporting the pier, put on the requisite amount of pressure, 42 pounds, and again broke the line.

Dr. E. W. Gudger, the famous bibliographer of fishes in the American Museum of Natural History, once wrote a paper entitled *Some Instances of Supposed Sympathy among Fishes.* In his search through the fish literature of the past hundred years Dr. Gudger came across a dozen or so examples of actions which, had they occurred between human beings, would have been interpreted as a sense of sympathy or an attempt to help a fellow-being in trouble.

Here are two of the examples cited by Gudger, together with an experience which occurred to him personally.

A man kept some turtles in an aquarium and he used to feed them on gudgeon. The turtles used carefully to stalk the gudgeon before making the final plunge and snapping them up.

One day, while an observer was looking at the aquarium, he saw a turtle stalking a small fish in this way. As he watched, he saw a larger and older gudgeon swim rapidly to the turtle and squeeze itself between it and its prospective meal. The rescuing gudgeon then covered the smaller one with its own body and, while splashing violently in the water, pushed the young fish into a more open part of the aquarium and thus out of danger.

The other story concerns two salmon. Hewitt says he once hooked a salmon and, as he played it, he noticed that other salmon in the same water seemed to be much excited. Then one of them swam alongside the hooked fish for a time and "then seemed to make up his mind what was the matter. He swam ahead of the hooked fish and then made a complete turn like a somersault right at the nose of the hooked fish, his tail of course hit the leader and broke it off. The whole action took place within 30 feet and we could see it all plainly."

Incidentally, members of the swordfish family have been known to help hooked companions in a somewhat similar way. Grey says a sailfish swam up and bit the line with which he was playing another sailfish. Farrington says he has known a marlin to swim for hours beside a hooked companion. Donne says a New Zealand fisherman observed a swordfish swim and leap out of the water side by side with a hooked fish for several miles.

Gudger's own experience also related to an apparent attempt by other fish to rescue the fish he had on his line. He hooked a large barracuda and, as he brought it to the surface, he noticed that on each side of it there was a barracuda nearly as large as itself.

"These had their heads in the region of the right and left pectoral fins of the hooked fish. It looked as if they had laid hold behind on the breast fins of their captured friend and were helping him hold back. . . . It must be confessed that it did seem as if all three fish were pulling back on the line. When the captive was brought near the boat its two companions disappeared."

3

Danger! Flying Birds

FROM the earliest days of human flight there have been occasional collisions between birds and aeroplanes. The earliest reference I have been able to trace dates from about 1910. A man was flying at Long Beach, California, when a seagull hit his machine and got wedged between the fin and rudder, which became immovable. The aeroplane crashed and the pilot was killed. (Hammel and Turner.)

A number of collisions were recorded during the First World War, but it was not until the Second World War, during which vastly more aeroplanes were flown than ever before, that the problem assumed serious proportions. Today, birds are a world-wide hazard to flying. I have records of birds proving a danger to aircraft in the Arctic, Antarctic, in the Pacific Islands, the Americas, Africa, India, Europe, and at sea.

The total number of aeroplanes that have been damaged or totally destroyed through collisions with birds will never be known. It is almost certain that a number, perhaps a high proportion, of unsolved air disasters have been caused in this way.

Some indication of the number of these accidents may be found in a report in *Time* (November 6, 1944), which stated that U.S. airline pilots were then reporting collisions with birds at the rate of two a week. It is known that during the migration seasons the rate is higher. When the accidents to military aircraft throughout the world are added to these figures for American civil airlines, it is obvious that the bird *versus* aeroplane problem has today reached serious proportions. In fact, the *Time* report says: "The bird-bumping problem is becoming so troublesome that airlines rate the Civil Aeronautics Administration's wind-shield-strengthening experiments as the most urgent present research project."

It is of interest to note that on occasion creatures other than birds have damaged aeroplanes. When Ernst Udet and a companion were flying low over the Serengetti plains of Africa to photograph a pride of lions, one of them leapt and badly clawed one of the aircraft (quoted by Supf). Beryl Markham, who used to spot elephant herds from the air for hunters in Africa, says that she has had big bull elephants charge at her aeroplane and challenge the "great bird" to give battle. A flock of bats once caused a pilot to make a forced landing. During a wave-hopping flight a pilot said a fish struck his propeller. An earwig once spoilt a bombing mission. It got into the bomb-releasing mechanism and prevented it from working, thus making the pilot bring his bomb-load home again.

While the larger birds are the greatest danger, quite small birds can cause appreciable damage. A sparrow hit the wind-shield of a big army observation aeroplane, crashed through it, and caused the pilot to make an emergency landing on a highway. Another small bird penetrated the wind-shield of a transport aircraft, passed through the bulkhead, travelled the length of the cabin, penetrated the rear cabin wall, and ended by lodging in the baggage department.

When collisions occur with flocks of large birds, serious damage to both aircraft and occupants may be inflicted.

An aeroplane flying at night at eight thousand feet ran into a flock of five cranes. One of the cranes was struck by the leading edge of one of the wings and embedded itself in it.

This aeroplane suffered very minor damage compared with another machine which collided at night with a flock of swans. A. L. Morse (1942), of the U.S. Civil Aeronautics Administration, who has taken special interest in the bird *versus* aeroplane problem, reporting on this accident, says:

> One swan penetrated the leading edge of the left wing; the second almost tore off the left vertical stabilizer, jamming the rudders, the third swan struck and dented the engine cowl, and two swans went through the propeller. A portion of a swan, taken from the wing after landing, weighed eleven and a half pounds.

During a bombing mission from England in September, 1942, some of the machines ran into a large flock of wildfowl. The damage they caused was so serious that several of the aircraft had to return to base, their bombing mission unaccomplished. A Spitfire, flying at high speed through a flock of starlings, was seen to dive vertically into the earth. There is little doubt that the pilot had been knocked unconscious by one or more of the birds crashing into the cockpit.

But it is not necessary for aeroplanes to collide with flocks of birds before they are severely damaged. A wild duck, which crashed into the cockpit of an R.C.A.F. flying-boat, hit the pilot in the face, stunned him and caused the machine to dive into the sea.

Another duck pierced the wind-shield of a transport aeroplane in Iowa, knocked out the pilot and a bad crash was only narrowly averted. An R.A.F. aircraft that was taking-off from the Lira aerodrome, in Africa, struck two white storks. The aeroplane crashed, burst into flames and was completely burnt out. (Pitman.)

De Labilliere says that a friend of his was flying across the Bristol Channel, when a seagull hit the machine and knocked the pilot unconscious. The aeroplane plunged into the sea, fortunately near a fishing vessel, whose crew rescued the pilot.

The force of the impact in some of these bird and aeroplane collisions is surprisingly high. In a report in the London *Daily Mail* for March 28, 1944 (late edition), it was stated that a bird had crashed through the wind-shield of a Spitfire. And such a shield is bullet-proof and at least one-and-a-half inches thick!

A still more vivid illustration of the force of impact is given in an experience of de Labilliere's. While travelling at about 400 miles an hour he struck a duck. He writes:

> After a safe landing, it was amazing to discover that the only visual damage to the leading edge of the port wing was a circular hole about the size of a duck's head. I might explain here that, in order to knock a hole through this very tough section of a Spitfire's wing, a heavy hammer and chisel would have to be employed.
>
> In due course the panelling of the wing was removed, and the complete carcass of the duck was recovered—not intact, of course, but in the form of bones, feathers and flesh, all compressed like a raw rissole. It was hard to believe that a four-pound duck could be forced, body, soul and spirit, through a comparatively minute hole such as its head had made. The statistics relating to such a collision show interesting results. Briefly, the force acting while the duck struck the wing was the equivalent of five tons, after which the pressure exerted to force the bird through the metal was almost two tons to the square inch, presuming that the bird only weighed four pounds.

The hedge-hopping technique, adopted by our pilots during one phase of the operations against Occupied

Europe during the Second World War, was responsible for a sudden increase in bird *versus* aeroplane encounters. When the crew of one bomber entered the interrogation room after a raid, they all had feathers in their caps. These decorations came from a covey of partridge which struck the aircraft over France. During the same raid a bird crashed through the wind-shield of another aeroplane. The pilot said: "My front gunner's turret was filled with feathers and the hole in the windshield let in an awful draught."

During the famous Eindhoven raid in December, 1942, when the Philips radio works were bombed, the chief danger to our aircraft appears to have come from the birds. A pilot who took part in the raid said afterwards: "By far the biggest danger was from the birds. We ran into flocks of lapwings, swallows, geese, and even herons. One of the fellows came back with a goose foot still on the trailing edge; another shocked his ground crew by climbing out of the cockpit covered with blood. It was from a slaughtered seagull."

After an earlier raid a machine came back with two seagulls embedded in its radiator. The gag at the aerodrome where this occurred was: "The Dutch will say tomorrow: 'Two of our seagulls failed to return.'"

Sea birds have proved a nuisance, if not a definite menace, to aircraft carriers. They perch on these ready-made islands and sometimes peck at the fabric on the wings of the waiting aircraft. Holes are thus made and the airworthiness of the machines may be affected.

During the First World War some French pilots used to take off with a bag of bricks in the cockpit. Their idea was to try to throw one of the bricks into their opponent's propeller. Two German aircraft were said to have been brought down in this way. The casualties among ground personnel from bricks that missed their objective are not stated.

An even more novel suggestion for air fighting emanated from France early in the same war. This was that eagles should be employed to attack enemy aircraft! In fact, six eagles were said to have been specially trained for the purpose. They were first accustomed to the noise of propellers and guns. Then pieces of meat were hung on model balloons in an attempt to get the eagles to rush at them—as, it was hoped, they would rush at enemy aircraft.

According to Wilkins, who recounts the story, an aeronautical journal in Paris, reporting on the eagle experiment, said:

> There is no airplane, and, above all, no dirigible, which could withstand such an attack. Given the rapidity of an eagle's flight, and the strength of its beak and claws, there can be no doubt that a company of properly trained eagles could annihilate, in a few seconds, the most powerfully equipped aerial fleet.

Wilkins adds that the French officers who trained the eagles did not subscribe to this view. Their idea was that if the birds could be trained to distract the crews of enemy aircraft, the machines might be brought down through the pilots losing control.

Fantastic as the whole idea may sound to our ears, it appears to be a fact, that of all birds, eagles are the most prone to attack aircraft. In this connection it should be remembered that eagles weigh up to fifteen pounds and in a stoop may touch 200 miles an hour.

The danger from eagles and other large birds was officially recognised by the British Air Ministry in 1934. A warning notice about them was issued to all pilots in the Near and Middle East. This notice said:

Only one rule can be given for the general guidance of pilots. Since the birds referred to invariably dive when alarmed, attempts to avoid them should be made not by endeavouring to pass beneath them, but by changing course.

A somewhat similar warning was issued to all pilots in the India-Burma theatre during the Second World War. Bramley writes:

> The records of the Army Air Forces, A.T.C. and R.A.F. in India-Burma are filled with instances of birds causing serious damage to aircraft. Every airfield has its flocks of birds, who seem to take special delight in hovering off the ends of runways. During some periods more aircraft have been rendered inoperative by birds than by Jap action.

Chapin was sent specially to Ascension Island in an endeavour to disperse the many thousands of sooty terns which persisted in nesting near the end of the runway on the island. He succeeded by having the eggs (some 40,000!) systematically smashed, which caused most of the terns to desert such an inhospitable area.

It was in this area that the suggestion was put forward officially that some pilots should be detailed deliberately to seek out vultures and make a flying attack at them in hopes of inculcating a healthy fear of aeroplanes. The aeroplanes so used were to be specially strengthened in the leading edge of the wing to resist collisions.

At about the time the Air Ministry notice was issued in 1934, an R.A.F. officer from India carried out some experiments at the London Zoo, with a view to finding an efficient "eagle scarer." Various whistles and other noises were made in front of the eagles' aviaries to see which produced the most violent reactions.

All the instances which have been reported of eagles

attacking aeroplanes may not, of course, have been deliberate—a number, perhaps the majority, may have been a result of panic. But there are some instances which appear to have been deliberate attacks. De Labilliere says that when he was flying in Scotland, two eagles made deliberate attempts to drive him away. Had he not turned aside they would have struck his machine.

During the Italo-Abyssinian War a Fiat fighter was attacked by an eagle. Although the pilot fired his machine-gun at the diving bird, it kept on and smashed the windshield, struck the pilot on the head and caused the aircraft to crash. A passenger aeroplane in Germany was struck on the side by an eagle which flew deliberately at the machine. During an eagle hunt from an aeroplane, organised by farmers in Texas to protect their lambs, at least one of the birds attacked the aircraft from which shots were coming. (Makin.)

The most remarkable instance of eagles attacking an aeroplane is reported by Day. The machine was a three-motored all-steel passenger aeroplane owned by Prince George Bibesco. When it was flying near Allahabad two eagles stooped on it. "The first flew straight into the middle engine, while the second dived from 10,000 feet and went through the steel wing like a stone, ripping a great hole." The aeroplane crashed.

Eagles are not the only birds which have been reported to make deliberate attacks upon aeroplanes. Walpole-Bond says: "I once saw a hobby stoop viciously time after time at an aeroplane." Makin says condors have attacked aircraft flying over the Andes. Williams says that during the summer of 1919 a pilot used regularly to fly low over a hayfield and, as he did so, a kingbird made a habit of dashing savagely at the aeroplane until it was out of reach.

Gladstone, writing of the First World War, says:

> Jackdaws were observed, in a French town, to leave their homes in the steeples and throw themselves upon aeroplanes, clinging to them and attacking them with their beaks as if to drive away these gigantic and unknown birds of prey.

A glider over Whipsnade Zoo was violently attacked by a macaw. Screaming furiously, the bird assailed the pilot with beak and claws and kept it up for a quarter of a mile. Similarly, a sailplane that flew in the vicinity of a raven's nest was viciously and repeatedly attacked by the male bird. (*Sailplane*, May, 1936.)

These are, of course, exceptional cases. It seems to be the general opinion of pilots today that birds are very little disturbed by aircraft when both are in flight and when the birds are not being harassed in any way. If an aeroplane approaches them closely, they usually wheel leisurely out of the way. When aeroplanes flew at slower speeds there were several reports of birds flying alongside the pilot, looking into the cockpit.

It is of interest to note that balloons appear to create far greater consternation among birds than aeroplanes. One balloonist says that most birds seem to go mad when the silent shadow of a balloon falls on them. Pigeons fly wildly to and fro in swarms and fowls cackle and make for the nearest cover. (Supf.)

Lack records the following remarkable reaction of an ostrich. He writes: "When in an Imperial Airways machine over the Kenya Game Reserve, we on several occasions flew close to a male ostrich, at which the latter would go down in the sand, spread its white plumes, and rock gently from side to side in display at the aeroplane."

Collisions with birds have occurred at considerable

heights. Several have occurred at about 13,000 feet. A pilot, while crossing the Andes at a height of 17,000 feet, collided with—or was deliberately attacked by—a condor. Other pilots have reported seeing storks and cranes flying at 20,000 feet. (Curtin.)

The main peacetime menace to aeroplanes from these collisions occurs during the migration seasons. The danger is especially acute where the centuries-old flyways used by the birds cross the commercial airways. As most migrating birds fly at night (they eat and rest during the day) bird and pilot are less likely to see each other in time to avoid collisions.

Pat Curtin, an experienced pilot of American Airlines, who is especially interested in this problem, makes the following suggestions for lessening the risks during the migration seasons:

> The altitude the birds operate at appears to be the level offering the most favourable wind for the direction of migration. With this in mind, it might be advisable in planning a flight to select an altitude other than that offering the most favourable winds for the particular seasonal direction of migration. It would be a simple precaution, and one that might pay dividends, especially at night.
>
> Just how successful a suggestion of this nature might be, could be proved by the information with which the airline pilots supplemented it. If, for one migration season, all the airlines asked their pilots to report all migratory flights of birds, and include the altitude they were observed at, the area they were over, estimated wind direction at that altitude, and the approximate direction the birds were headed at the time, it might be possible to learn enough about the habits of the birds to reduce the number of bird strikes on aircraft by more than half.

Radiolocation, or radar, may help to lessen the number of bird *versus* aeroplane collisions. It has already been proved that radar can detect birds in flight. When a herring gull was suspended from a captive balloon it was found that the echo from the bird was clearly separable on the radar screen from that given by the balloon. (Lack and Varley, and Buss.)

Birds have been responsible for both invasion and air-raid false alarms. In a letter, David Lack, who worked with Dr. G. C. Varley on radar during the Second World War, tells me that birds were a great nuisance to radar operators. He and Varley had to give special talks to the R.A.F. operators on the coastal chain on how to distinguish birds from aeroplanes on the radar screen.

When this country entered the war, radar operators had the same trouble. On June 11, 1943, a large flock of pelicans was detected on the radar screen of stations on the west coast, and resulted in air-raid signals being sounded in San Francisco! (See *International Correspondence on Aviation,* Geneva, No. 1044, August 25, 1945.)

Experiments have been carried out with a view to fitting a collision warning device, using radar, to the instrument panels of aeroplanes. (Brooks.) Another suggested use for radar is as an aid in plotting the migration routes of birds. The results may help the airlines to re-route some of their airways during the migrating seasons to avoid colliding with the main stream of migrating birds. In Britain trained peregrine falcons have been used in attempts to keep airfields clear of birds. (Lane.)

In passing, although aeroplanes have undoubtedly killed many thousands of birds during the migration seasons, there was one occasion when they were responsible for saving many birds. In the autumn of 1931 a blizzard which swept over Austria prevented thousands of swallows from migrating south, with the result that they were dying from want of food. Bird-lovers were instrumental in getting two aeroplanes to *fly* some of the swallows on their journey! Twenty-seven thousand swallows were collected in and around Vienna, put into the aeroplanes and then flown across the Alps to Venice, where they were liberated under a sunny sky in a temperature forty-two degrees F. above that of Vienna. (Beadnell.)

Captain Neil T. McMillan, an experienced civil airlines pilot in this country, makes an interesting observation about these encounters between birds and aeroplanes. Although he says he has killed many birds while flying, his experience leads him to believe that many small birds are carried over the wings of the aeroplane by the airflow. Nearly all the birds he has killed were struck, either by the nose of the machine or the under-side of the fuselage. In fact, he believes that when an aeroplane is flying at full speed the airflow will carry any bird over the wings (quoted by Terres). But it will be obvious from some of the incidents quoted earlier in this chapter that some birds have undoubtedly been hit when aeroplanes were travelling at high speed.

Although aeroplanes may be struck on almost any part, the wind-shield appears to be the place where collisions occur most frequently. (Eiserer.) This is, of course, the most vulnerable place, because the pilot is immediately behind. Aeronautical engineers have, therefore, devoted a good deal of time and ingenuity in endeavours to provide aircraft with bird-proof wind-shields.

The Bureau of Standards, the National Advisory Committee for Aeronautics, and the Civil Aeronautics Administration, have been chiefly responsible for this work. Towards the end of the Second World War, the British Ministry of Aircraft Production conducted its own investigations into the subject. I had the privilege of playing a minor part in the Ministry's investigations by supplying information on the speed, weight and height of flight of birds likely to be encountered by pilots flying over and around the British Isles.

From the point of view of the investigators, a bird was a "liquid object in a fragile container," which annihilated itself at high speed against the glass. In the earlier experiments, a standard bird was assumed to weigh about four pounds and to collide with the wind-shield at a relative speed of 270 miles an hour—200 for the aeroplane and 70 for the bird. Such an impact would mean that about 10,000 foot-pounds of energy would have to be dissipated by the wind-shield. If a bird weighing 20 pounds, such as a swan, crashed against it at the same speed, there would be a pressure of over 100 pounds to the square inch on an average commercial aircraft's wind-shield, which means that a total energy of approximately 50,000 foot-pounds would be expended against it. It is not surprising that the wind-shields in use when the experiments started were frequently shattered by birds.

The investigators were faced with the initial difficulty of simulating the actual conditions of the bird collisions. Objects of various weights were first made up to take the place of birds. Among these were tennis balls partially filled with lead shot and weighing one pound; a dart weighing three and three-quarter pounds, with a solid rubber ball at the striking end; a vulcanized cylindrical projectile, wrapped in cellophane, weighing half a pound;

projectiles made by moulding one-inch sponge rubber around a core consisting of a small rubber ball filled with sufficient lead shot to make the total weight three or four pounds as desired. To simulate the smaller birds, tomatoes and paper bags filled with water were used. These "birds" were propelled from a two-inch air cannon or from a five-inch A.A. gun at various test wind-shields.

The more recent tests, carried out at the Westinghouse Electric and Manufacturing Company's plant at East Pittsburgh, Pa., have all been made with chickens and turkeys weighing from three to seventeen pounds. The birds were painlessly electrocuted immediately before the tests and then put into a light cloth bag. It was considered that for test purposes a seventeen-pound turkey would simulate a collision with a twenty-pound swan (with the exception of condors, swans are the heaviest birds ever likely to be encountered) as the wings of the swan would extend beyond the boundaries of the wind-shield.

A compressed-air gun was used to project the bird carcasses at the test wind-shields. Two interchangeable barrels, five and eight inches in diameter and twenty feet long, were used. The speed of projection could be varied between 50 and 450 miles an hour. The dynamics of the impact between the carcasses and the wind-shields were studied by the use of a 35 mm. high-speed motion picture camera taking 1,500 pictures per second. Various electrically operated gauges and oscillographs gave additional data on the tests.

These American experiments have provided valuable information on the problem of bird collisions with aircraft. But as they involve shooting bird carcasses at stationary wind-shields, they do not duplicate exactly the conditions under which actual collisions take place. There is, for example, no allowance for wind resistance against the wind-shield, such as would be experienced in an aeroplane. The experiments dealing with this problem which have been carried out in England have sought to reproduce as exactly as possible the actual conditions which occur when an aeroplane and a bird collide.

Through the courtesy of the officials engaged in making the experiments at the Royal Aircraft Establishment I was able to witness some of this work.

In one experiment a dead rook weighing about one pound was suspended by easily broken threads over a track on which was a small rocket-propelled trolley carrying part of a fuselage, including the wind-shield (see plate 18). The rook was in line with the centre of this wind-shield, which was not of the kind specially designed to be proof against collisions with birds.

After the trolley was wheeled to the end of the track, eight three-inch rockets were fitted at the rear. A high-speed motion picture camera, taking 200 pictures per second, was focussed on the rook. At a given signal the camera was started and the rockets fired electrically. There were spurts of vivid orange-red flame from the rear of the trolley and with a roar it hurtled down the track. The wind-shield hit the rook, parts of which shot into the air.

The impact took place at a speed of 234 miles an hour. The electrical timing device at the side of the track showed that this speed was reached within 50 yards of the start. The wind-shield, when it was examined afterwards, was found to be shattered and the rook was found near the track badly mutilated. The high-speed motion picture films showed that the bird was impelled in the direction corresponding to the angle at which the wind-shield was set in the fuselage.

The various experiments which have been carried out in this country have proved that the conventional quarter-inch safety glass wind-shield collapses under the impact of four-pound birds projected at a velocity of 75 miles an hour, which is less than the normal landing speed of modern commercial aircraft.

The tests also revealed that even when no penetration of the wind-shield occurred, its transparency, and therefore the pilot's vision, was destroyed through the crazing of the glass face, and large quantities of dangerous glass splinters were thrown off the rear face of the shield at a speed of about 350 miles an hour.

After a number of tests with various types of experimental wind-shields, it was found that one made of laminated glass-vinyl, with extended plastic edges and other strengthening devices, having a total thickness of about three-quarters of an inch, would resist the impact of four-pound carcasses projected at 300 miles an hour and a fifteen-pound carcase at 200 miles an hour.

Incidentally, the British Air Registration Board, the controlling body for civil aircraft in England, specifies that wind-shields should be of sufficient strength to withstand the impact of a four-pound bird "when the aeroplane is flying at the speed appropriate to climb immediately after take-off."

Other devices designed to lessen the bird hazard to aeroplanes which have been considered are: the installation of two separate panes of glass, one behind the other; the use of smaller panes in the wind-shield with more strength per unit (further consideration has indicated, however, that within limits a larger pane, being more resilient, is more bird-proof); installing a metal grating which can be dropped over the wind-shield; and the use of a protecting shield for the pilot which will cut up birds before they crash into him.

4

Animal Accidents

THE Right Hon. L. S. Amery, the English politician, tells the story of a bear that came across a cache of food left by some travellers in Canada. Among these stores was a 20-pound sack of dried apple chips, which the bear eagerly devoured. Later it went to a nearby stream and drank a gallon or so of water. The effect on the 20 pounds of apple chips may be imagined. When the travellers returned they found the bear by the stream, split from stem to stern!

A similar cause of death has been recorded for the recently extinct passenger pigeon. These birds often raided fields which had just been sown with peas. Then, with their crops full, they flew to a stream and drank. Soon the water caused the dried peas to swell, and they either burst the pigeons' crops or choked them. (Mitchell.)

These are two examples of the many ways in which food can cause death. Sometimes an animal will "catch a Tartar," and be killed by its intended prey: both frequently dying. Thus ospreys and eagles occasionally sink their talons into fish too heavy to lift and are consequently drowned, the fish probably dying later. Stoats and weasels, when seized in the talons of birds of prey and taken aloft, occasionally fix their needle-like teeth in their would-be captor and both crash to their death.

Fishes occasionally meet their death through attempting to swallow fellow fishes too big to pass down their throats.

Gudger (1929) details and pictures a number of such instances. A lake trout, 8¼ inches long, was found dead with a saw-belly, 5 inches long, firmly wedged in its mouth. A 44-pound striped bass was choked when it attempted to swallow a 2-pound carp.

Two pike are sometimes found dead with the head of one firmly wedged in the mouth of the other. Not all such incidents are the result of attempted cannibalism, although this may well account for some of them. Cholmondeley-Pennell says two pike were kept in an aquarium. When a piece of food was thrown about midway between them, they simultaneously rushed at it, with the result that the head of the smaller fish entered the open mouth of the larger. Here it became so firmly fixed that it was an appreciable time before the two fish managed to free themselves.

While all classes of animals occasionally die through attempting to swallow prey too big for them, this cause of mortality appears to be most rife among birds. The prey gets wedged half-way down the throat and the bird can neither swallow nor eject it. Sometimes the prey fights back.

Several instances have been reported of large eels thus killing herons. When seized, the eel wraps its free end round its captor's long neck and strangles it. Hudson says he once watched a cormorant nearly suffer the same fate, but it managed to free itself from the eel just in time. A kingfisher that once seized a 6-inch long eel had its neck broken when the eel coiled round it. Cuming says he had a merganser sent to him which had an elver 2½ inches long tightly plugged in its left nostril. The other end of the elver was down the merganser's throat.

Prey that has been swallowed alive occasionally fights back from inside the body of its captor. McAtee (quoted by Terres) says a whip-poor-will once swallowed alive a large and powerful beetle, which burrowed directly through the bird's gullet. Ross says a black phoebe was killed by a honey-bee stinging it in the roof of its mouth.

Gudger (1922) has written at length of the foreign bodies, including prey, which have been found embedded in the tissues of fish. One of the most remarkable instances concerns a cod which swallowed a live hermit crab. Once inside the fish, the crab went exploring and, penetrating the wall of the cod's stomach, it passed into the body cavity where it was found, mummified, by the man who cut up the cod.

Another cod was found to have a curiously shaped knife, with a brass handle, embedded in its flesh—with the blade closed. But, judging by the condition of the fish which harbored such strange guests, accidents of this nature are by no means always fatal.

Poisonous and inedible food, of course, kills many wild creatures. Birds die through eating such things as phonograph needles, nails, fish-hooks, glass, and other highly indigestible fare, which penetrates the digestive system and produces severe pathological conditions. Impaction of the intestine may occur through swallowing coarse vegetable fibers, flexible wire or string. Ducks die through swallowing lead shot. Some birds die of phosphorus poisoning by feeding on the remains of fire-works. (Stählin, Morgan, and Austin.)

The speed at which birds travel is responsible for a fair number of accidents. Macintyre says an exhausted gannet was washed ashore on the coast of Mull, Scotland, with the body of a guillemot round its neck. Evidently the gannet, in one of its characteristic headlong dives into the sea for a fish, had accidentally speared the guillemot.

Macintyre says, should two gannets "plunge for the same fish at crossing angles, one or both will utter a warning cry, or cries, and the pair, although falling like plummets through the air, 'balk' their dives, sheering aside with what sounds like soldiers' language. A danger is that a swift diver, like the guillemot or the puffin, may make an under-water attack on a fish which it has selected and is plunging at."

Thomson says that very occasionally a gannet may run its beak into the open mouth of a fish. In such an accident the bird would almost certainly be drowned and the fish killed. One gannet dived on a garfish, and the sharp upper mandible of the fish passed obliquely through the bird's eye and pierced the brain. Gurney, who records this incident, says: "Some species of birds are more singled out for misadventure than others, and I really think this can be said of the gannet."

On rare occasions birds collide in mid-air and sometimes die as a result. Joy saw a swift fly out of a railway goods shed and collide with another swift just outside. Both birds fell and died about five minutes later.

Writing in *The Field* for December 8, 1945, E. Cobden says he has twice seen birds collide fatally: once when two ducks collided over a playing field and one of them was picked up dead, and again when two cormorants collided high in the air and one fell.

Protheroe mentions a remarkable accident of this type when a pheasant flew across a flock of manoeuvering starlings and was killed, presumably by being struck repeatedly before it could get out of the way. Other aerial collisions are mentioned by Gladstone.

Very light obstacles suffice to damage a bird's wing. Falcons, swallows and swifts have damaged or even broken their wings by striking twigs or, as has happened occasionally, fishing rods.

Kearton says birds have been known to break their wings in mid-air. In one instance, an old road-mender at work behind a line of butts saw a covey of grouse coming towards him and playfully putting the head of his long-shafted hammer to his shoulder, he drew a bead on the foremost birds. One bird, evidently thinking the old man had marked it down, suddenly twisted in its flight and then, in the most orthodox manner, fell with its wing broken! Presumably the bird had been shot at previously and a pellet or pellets had damaged the bone. The sudden extra exertion occasioned by the violent evasive action did the rest.

Kearton appears to think that, on occasion, the mere exertion of flight will break an undamaged wing. He says a sportsman flushed an owl while out quail shooting in Egypt. He playfully raised his gun, when the bird twisted in flight and fell to the ground. Upon examining it, the sportsman found it had broken its wing. Gladstone records a similar instance of which he was a witness. A sportsman pointed an empty gun at a covey of partridge and one of them, presumably frightened, twisted in flight, fell to the ground with a broken wing and was caught.

Birds of prey, wonderfully skilful on the wing as they are, occasionally meet their death when, in violent pursuit of their prey, they collide with various obstacles.

But such accidents are by no means confined to predatory birds, and the records of ornithology contain a number of accounts of birds of other species which have died in collisions with various obstacles. Structures erected by man, dealt with later in this chapter, account for many of these deaths, but natural obstacles also take their toll.

Shadle (1931) found a dead crow hanging in a maple tree. On examination he discovered that a sharp stub of a dead branch had penetrated the crow's left wing, presumably on a downward beat. Shadle said in one place the whole surface of the branch had been torn away, mute evidence of the crow's desperate efforts to free itself.

Forbush says he has seen a ruffed grouse fly at great speed into a wood, strike the branch of a tree and fall to the ground. Apart from the shock the bird appeared unhurt. Forbush adds that he had the body of another grouse sent to him that had collided with the forked end of a dead branch. One of the prongs had driven three inches into its breast and another had torn the head and neck from its body.

Autopsies on other grouse have revealed some remarkable accidents which they have suffered and yet have survived. Allen (quoted by Terres) says he has found a fair-sized twig encased in membranes in a grouse's body, with no apparent inconvenience to the bird. Apparently the twig had been forced down the bird's throat when it was flying at high speed through undergrowth.

Another grouse had apparently had an accident similar to the first one quoted by Forbush, for part of its crop, containing acorns, had, by some terrific buffet, been torn away and pushed under the skin of the lower breast. The crop itself had healed perfectly.

Even more remarkable are the rare examples of birds living with twigs or stakes completely transfixing their bodies. There exists a specimen of a passenger pigeon with a nine-inch-long beech twig sticking through its body. The twig had pierced the bird from below and had come out of its back a distance of some four inches. It is thought that the pigeon fell out of its nest when a squab and had been impaled on the twig. When collected the bird was old and the twig was much worn. (Terres.)

Charles K. Nichols has reported a similar injury to a robin and supported his account with a photograph of the living bird (see plate 23). The bird used to come regularly to the garden of Dr. M. E. Wigham, in New Jersey, where Nichols made a detailed study of it through binoculars. He writes:

> The stick appeared to enter the back at the left of the backbone and behind the heart and the lungs, penetrating the body in the area of the stomach and kidneys, but just enough to one side to miss them. It would seem that the stick must lie against the left peritoneal wall, as it could hardly go anywhere else without damaging a vital organ. The stick described approximately a right-angle with the backbone, and came out of the breast probably through the lower ribs. The projection from the breast seemed to be about a quarter of an inch closer to the median line than at its point of entry in the back. About two inches of the stick projected from the back of the bird and about an inch protruded from the breast. The stick was a straight twig about three-sixteenths of an inch in diameter and had a slightly roughened bark attached.

The robin behaved normally, except that it was naturally somewhat awkward on the ground and in flight. It mated twice in the season it was under observation and took a normal part in feeding the young and in protecting its territory from intruding birds. The bird returned some two years after it had first been seen, and the stick appeared a little shorter and the ends were frayed. It disappeared finally a few weeks later.

G. L. Bates shot a hawk in the Cameroons, and on examining it he found part of a small native arrow that had pierced one eye and part of its head (quoted by Terres).

Vegetation of various kinds is occasionally responsible for some strange accidents to birds. Cuming says he saw a number of birds fly up from a field of peas, but one of them, a linnet, almost immediately fluttered to the ground again. One of its legs had been caught so firmly by a tendril that it would probably have died of starvation had it not been rescued. Osmaston saw a scops owl collide with a branch of a prickly climber. The owl was caught by its eyelids, by "at least two thorns both above and below the eye," so securely that it was doubtful if it could have escaped unaided.

E. T. Booth (quoted by Kearton) thus describes an unusual accident which befell a kingfisher. While at Hickling Broad in Suffolk, England, he noticed the kingfisher splashing in the water at the side of a dyke.

> The bird had evidently at some former time been struck by a shot which had passed through the upper mandible. This wound was quite healed up, but a small piece of horny substance of the beak had been splintered, and into the crack produced by the fracture two or three of the fine fibres, which form part of the flowers or seeds of the reed, were so firmly fixed that the bird was held fast. It must have been flying up the dyke and brushing too closely to the reeds that grow on the banks, and been caught in the manner described.

Rogers records another strange accident to a species of Indian kingfisher that was found lying helpless with its feathers and wings thoroughly stuck together by the gluey seeds of *Pisonia excelsa*. In the Keeling Islands these seeds have been known to cling to the feathers of herons in such quantities as to cripple them and sometimes to cause their death. McNeil says hundreds of terns and mutton birds, on Nor'west Isle in the Great Barrier Reef, Australia, die each nesting season through getting their wings and bodies entangled with bunches of the extremely sticky pisonia seeds. Joseph Janiec, in a letter to Bent (1940) says he found a humming-bird firmly stuck by its stomach feathers to the prickly, pointed stamens of a pasture thistle. Another of these tiny birds was found caught by its neck between two gladiolus stems. (*Nature Magazine*, June, 1935.)

In their search for food birds occasionally collect strange impedimenta. Sutton says he found a pine warbler in such a weakened condition that he killed it. He continues:

> Examining the specimen closely, we discovered that the right foot was almost completely encircled by a piece of pine bark, which refused to come loose and which, judging from its weathered appearance, had been detached from the tree for some time. The bird was very thin and its plumage badly worn. Most of the feathers of the crown were missing. It is our belief that the unfortunate bird had caught its foot in the bark, while searching for food, was held captive for some time, finally managed to break free from the tree, but was never thereafter, because of the considerable burden, able to get about normally.

David G. Nichols reports finding a varied thrush which was apparently unable to fly. On examination it was found to have on its bill a large acorn, a sharp corner of which was forced into one nostril. The bird was very weak and must have been partially starved. When the acorn was cut free with a pair of scissors the thrush flew feebly away.

Cyril Newberry has sent me the photograph (plate 25) of a great spotted woodpecker which was found dead with an oak gall firmly fixed on its beak. Considerable force was required to remove the gall. Incidentally, Hall reports finding a woodpecker dead with its beak embedded in the soft wood of a Wellingtonia tree. Had the woodpecker struck so hard that its beak had become fast and it was unable to withdraw it?

Hairs and strings are another menace to bird life. A correspondent of Frances Pitt related that two starlings fell fluttering to his feet. They had become tied together by a piece of string. Kearton says a small bird became entangled in some of the long hairs of a colt's tail and, unable to free itself, was dragged and bumped about a field until it died. Incidentally, horses, sheep, and cattle, as well as mowing-machines, are a danger to ground-nesting birds.

It is during the nesting season that the greatest number of fatalities of this nature occur. Hairs, fibers and string used in making nests sometimes cause the death of the nest-makers. Evermann says a colony of over two hundred cliff swallows, nesting in a shed, plastered many horse-hairs into their nests. He found "some six or eight birds" hanged by these hairs. It is probable that at least an equal number of nestlings were similarly killed.

A nestling's leg can easily become trapped by a hair or fiber and the bird be held prisoner until it eventually dies of starvation. Several examples of this type of accident were reported in *British Birds* for September, October and November, 1930. In one instance a piece of wool had become twisted round the two little spines at the base of a young chaffinch's tongue. Low found a young coot fatally snared in a loop of reed grass which had been used to make the nest.

Sometimes hairs get mixed with food and, when this happens, occasionally two birds each have an end of the hair in their crops and thus become joined together. Witherby reported such an accident, in which both birds were picked up dead.

Water birds sometimes become entrapped in aquatic vegetation and are drowned before they can free themselves. Kearton records two such fatal accidents which he witnessed.

Small fish and other water animals also have their troubles from under-water vegetation. Conspicuous among such foes is the bladderwort, a submerged water plant equipped with tiny bladders. Sophia Prior thus describes how these trap small animals:

> There is a valve on the posterior free edge lined with numerous glands, each consisting of an oblong bead and a pedicel. This valve opens only inwards and is highly elastic. Small animals can enter the bladder through this valve, which shuts instantly behind them and does not yield to pressure from within, so that it is impossible for an animal to escape once it is caught in this prison. Under favourable circumstances many of the bladders may be found to hold as many as eight minute crustacea.
>
> A bladderwort has been described by Moseley which entraps young fish and spawn. "Most are caught by the head, and when this is the case the head is usually pushed as far into the bladder as possible till the snout touches the hinder wall. The two dark black eyes of the fish then show out conspicuously through the wall of the bladder."

It has been suggested that bladderwort should be used in mosquito control, presumably to dispose of larvae.

Prior says a small frog has been trapped by Venus's fly-trap. She says also that birds and small mammals are occasionally drowned in the pitchers of the large pitcher plant (*Nepenthes Rajah*) of Borneo. These pitchers are sometimes a foot across and contain nearly a gallon of water.

Apart from plants, such as the Venus's fly-trap, which are specially adapted for capturing insect prey, other forms of vegetation are a danger to insect life. The spear-like tips of marram grass sometimes impale insects, and Stelfox recounts seeing such an impalement take place on a windy day. He writes:

A large fly, resembling a bluebottle, was flying upwind in front of me and attempted to alight on the tip of one of the nodding blades of the grass, but at the last moment this sprang away from it and instantly returned and speared the fly.

Stelfox adds that he has seen a bee impaled on this grass, about an inch of which projected through and beyond its body.

Bees are occasionally killed by the flowers of the torch lily, popularly known as the red-hot poker. A writer, giving the initials "C. Q.," published a letter, and a photograph supporting his evidence, in *Country Life* for October 21, 1916, saying he found hundreds of dead bees enclosed in the long tubular flowers of one plant. It is thought that the bees enter the flowers in search of nectar about the time they are fading and losing their elasticity. The bees, having forced their way in, are held fast by the natural contraction of the fading flowers and are unable to withdraw.

Williams says when he was in Ecuador he noticed many insects sticking to the adhesive burr-like seeds of the sand-burr grass. In one field alone he estimated that many thousands of insects must have been killed by this grass, including moths with a two-inch wing-spread. In Australia even powerful beetles have been thus caught.

The spines of the burrs are minutely barbed so that the seeds, catching in the coats of passing animals, will be widely distributed. But many insects are not strong enough to detach them from the parent plant and, even if they were, they would have little chance of getting rid of the seeds.

Bats are sometimes caught in a somewhat similar manner to insects, especially on the hooked burrs of burdock. Harlow photographed and described one such incident where a brown bat was found spreadeagled on the myriads of burdock head spines, and was dead by the time a would-be rescuer arrived. Stager records another bat found dead, fast to the spines of a large desert shrub, and Venables found a long-eared bat impaled by its fully-extended wings on the thorns of a rose bush.

Small ground mammals also have their tribulations from vegetation. Goin found a dead mouse impaled on a spine of a trailing prickly pear. Miller reports finding a ground squirrel which had three large cholla burrs stuck to its body. As the squirrel attempted to run, the burrs upset it and, as it rolled over, the spines were driven more deeply into its body. Other small mammals sometimes become entangled with grasses and stalks and are held prisoners until they die.

Even the larger mammals occasionally have serious trouble with their natural surroundings, especially with trees. Beavers are sometimes trapped and killed by the trees they have felled. It is said that a beaver always fells a tree so that it will fall in a desired direction, but this is not so; the beaver makes the deepest cut on the side most easily reached. (*Life,* October 20, 1943.) Two hunters in Northern Rhodesia heard one night a rending crash followed by an elephant's piercing scream. In the morning they found an elephant's tusk lying at the base of a tree which was split. The tusk was bloody but unbroken and appeared to have been completely drawn from the elephant's head. Evidently the elephant had smashed into the tree and its tusk had been driven in so deeply that when the great beast had backed away the tusk had been pulled from its socket. (Hubbard.) Twist records finding an elephant held fast by its head in the fork of a fallen tree.

I have two records of bears being trapped by trees. A black bear that had stepped on a fox trap and had it clamped firmly on one of its hind legs tried to climb a small maple tree. But, hampered as it was, this proved difficult and, when part-way up, the bear slipped and broke its neck in a crotch of the tree. (Luttringer.)

The other bear accident would also have proved fatal had not human aid been available. A bear found a squirrel's store of nuts in a hole in a tree. In its eagerness to get them the bear got its head stuck fast in the hole, and was released only when the hole was carefully enlarged with an axe. (Bridges.)

More unfortunate was the cow which strayed into a wood and swished her tail round a small tree. By some mishap the tail became firmly tied to the tree and the cow's vigorous efforts to free herself only made the knot faster. She was thus eventually found—dead of starvation. (*Public Safety,* quoted in *Science Digest,* February, 1940.)

Of all large mammals, deer appear to have the most trouble with trees. Jay C. Bruce (quoted by Seton) found a dead blacktail deer tightly wedged in the crotch between two trees. It had apparently been reaching up to eat mistletoe from the trunk of one of the trees, had slipped and one of its feet got caught. When found the deer lay belly up on the ground, its leg twisted to the shoulder, showing how frantically the poor beast had fought for its life.

Another deer was found dead, hanging by its neck from a large crack in a tree stump. Luttringer, who records this incident, tells of a still more unusual accident in which a deer was found dead of starvation, with the forepart of its head tightly wedged in an opening in the trunk of a large tree.

Luttringer says that the deer had evidently "been lured to thrust its head through the wide lower part of the opening by appetizing lichen and ferns growing in the dank soil along the inner base of the tree. Surprised or frightened while thus engaged, the sensitive creature did what it is his breed to do, raised his head with such whip-like energy and rapidity that it became wedged in the narrow upper portion of the trunk cavity." (See plate 27.)

Deer frequently travel at high speed through wooded country, and occasionally they spear themselves on bayonet-like pieces of projecting wood. Seton has several records of deer which, when skinned, have been found to have comparatively large pieces of wood in their bodies. R. Clarke Fisk (quoted by Seton) found, on skinning a whitetail deer he had shot, that it had a fir branch, over a foot long and more than half an inch thick, driven into its body.

> It had entered between the fourth and fifth ribs on the right side, missed the right lung, pierced the top of the diaphragm and the point of the liver, and rested against the under side of the backbone. That the animal met with this accident while it was yet young I am thoroughly convinced, for the end at the ribs had been entirely drawn into the opening of the heart and lungs, and had thoroughly healed on the outside. The skin, which I now have, shows only the faintest trace of a scar. There was not a particle of pus or inflammatory matter of any kind. In fact, the branch, covered, as it was, with white skin, exactly resembled one of the long bones of the leg. The animal was healthy and fat, and the meat was fine.

A hazard peculiar to male deer and antelope occurs during the rutting season, when the bucks engage in fierce duels. When fighting, these animals bow their heads and engage their horns. Sometimes the tines become so firmly interlocked that the most desperate efforts fail to free them and the deer die of exhaustion and starvation. The horns are sometimes so tightly locked that when the carcasses are found two men are unable to pull the horns apart. Occasionally three bucks get their horns interlocked and perish.

Seton says it is so common to find dead buck elk and Virginia deer with interlocked horns, that he believes one per cent. of them lose their lives from this cause. But he believes an even higher percentage of whitetail bucks die through this cause—owing probably to the peculiar formation of the tines.

A party hunting in a wood near Indian River, Michigan, came to a fairly open space and found the ground for nearly an acre torn and furrowed by the hooves of two whitetail bucks. Near the center of this space lay the animals themselves, their horns inextricably locked. One deer was dead and the eyes of the other were already glazing.

Although it does not happen so frequently with them, as their horns are differently shaped, moose bulls occasionally die through interlocked horns. The bleached and crumbled horns and skulls of two moose were found on examination to be locked together. A tine of one horn was driven into the eyesocket of the other skull and was embedded in the bone at the back. (Seton.)

I have read of two instances of kudu bulls, whose horns are spiralled, getting them interlocked while fighting and dying as a result. Pohl records one such instance, in which both bulls died. Yates says, when two other bulls were fighting "a sudden, mutual onset caused the tip of each right horn to glide instantaneously along the spirals of the corresponding horn of the adversary, with the result that the two became so firmly intertwined that all subsequent attempts to wrench them apart were unsuccessful." One bull died, but men came along and sawed off its horns in time to release the other kudu. Kipling says Indian blackbuck, which also have spirally-shaped horns, sometimes die through getting them interlocked.

Mammals are sometimes killed by colliding with obstacles when not looking where they are going. Kangaroos often look over their shoulder when travelling at high-speed and are sometimes killed before they know what hit them. (Fisher.) A cub killed itself in front of hounds by running into a stone wall and breaking its neck. Rabbits likewise are occasionally killed when they collide with obstacles they fail to see in their haste.

Hares often look behind them when running, doubtless because their chief danger lies in this direction. This habit sometimes has dire results. Two hares, equally oblivious of where they were going, collided at full tilt. They both died —one from a broken neck. Another hare collided with a dog and again the result was a broken neck—for the hare. And then there is the well-known story of the hare, hard pressed by hounds, that ran straight towards an old country lady. She stooped down, enfolded the hare in her lifted skirt, and despatched it with a couple of blows at the back of the neck! (Bryden.)

Some of the examples mentioned elsewhere in this chapter, of birds colliding with various obstacles, may also have been due to their not looking where they were flying. Macintyre says he found four grouse lying close together at the foot of a moorland fence, which had bushes tied at intervals of a yard along its whole length and should, therefore, have been clearly visible. On another occasion he saw a pack of grouse strike a fence and then he realised how such accidents happen. The leading birds in the flock see the fence, rise over it, drop, and continue flying at their former level, but the rearmost birds, flying blindly, crash into it.

Gladstone says a grouse has been known to kill itself by flying against the barrels of a gun. Cuming says a wild duck, flying down a glen, crashed into a bucket full of water, which a girl was carrying on her head. The bird was crushed to death and the bucket had a deep dent made in it.

Another group of fatalities which may be called accidental, are those where death has come through the agency of an animal or organism not normally regarded as an enemy of the species attacked. Thus, following the famous dictum of what constitutes news, if a bird bites a snail that is not news, but if a snail bites a bird that certainly is.

Lowe (1943) says, in Africa he has had many valuable trapped specimens completely ruined by snails. "This seems to occur at the snails' breeding season, a creature caught overnight being ruined. They eat a large patch of the skin completely and then devour the flesh." In a later communication on this subject, Lowe gives further evidence of snails eating birds and also mice. It is known that snails sometimes attach themselves to live birds, and Lowe (1944) says: "I should not be at all surprised to hear that birds sleeping on the ground are at times killed and eaten by snails." Similarly, leeches occasionally attack birds and other animals, and sometimes death may result. (Cunningham.)

Insects also take their toll of bird life. Browne records a remarkable attack by a mantis on a sunbird. The bird was hovering round a branch on which the mantis was resting and, whether in fright or otherwise, the mantis struck out at the bird with its spiked forelegs and then scalped it, dropping it dead. Browne adds: "Considering the way in which the mantis's forelegs are armed, and that it weighs considerably more than the bird, there is nothing inherently impossible in what occurred."

A large dragon-fly has been seen fastened on the back of

a ruby-throated humming-bird on the ground. The dragon-fly had seized the humming-bird by its neck and it is probable that the bird would have shared the fate of the sunbird had it not been rescued. (Bent, 1940.) Burr says the large bug, *Belostoma grande* and its relatives, some of which are over four inches long, sometimes attack small vertebrates, and one has even been known to kill a woodpecker while it was sitting on a tree. The bug pierced the woodpecker's skull and sucked out its brains. If carelessly handled these bugs can inflict a painful bite on a man.

Dr. James P. Chapin, Associate Curator of Birds at the American Museum of Natural History, told Terres that while in the Belgian Congo he shot a swallow for scientific purposes and, on picking it up, found that one leg had been bitten through by a driver ant. The head of the ant was still fastened to the swallow's leg, part of which had died and atrophied. A starling which was caught had a four-inch long centipede on its leg, which the myriapod had severely bitten. (*Victorian Naturalist*, August, 1944.)

Spiders are another enemy of birds. In fact, the South American spider *Mygale* owes its popular name, "the giant bird-catching spider," to this trait. Small species, such as humming-birds, appear to be fairly frequent victims of spiders. Some spiders weave their webs so that the surface lies in a horizontal plane, with the result that birds flying upwards from the ground are easily trapped.

Occasionally, comparatively large birds become enmeshed in spider's webs, some exotic specimens of which are much larger and stronger than any known in temperate climates. Often, of course, the bird is able to tear its way out of the web, but sometimes a bird as large as a robin will be caught and killed by the spider before it can escape. (Ealand, and Gudger, 1924 and 1931.)

That some spiders will deliberately attack birds has been proved more than once. Captain Thomas Sutton, in India, placed a young sparrow under a bell glass with a large spider (*Galeodes*). The moment the bird moved the spider seized it by the thigh. Then it transferred its hold to the sparrow's throat and quickly killed it. C. L. Doleschall, in the Netherland East Indies, put a giant Javanese spider (*Mygale javanica*) into a box and then introduced a rice bird. "At once the spider sprang on it, embraced it with its feet and sank its poison-bearing fangs deep into the region of the spinal column. Within thirty seconds the bird died in a tetanic spasm. The spider at once began sucking the juices of its prey." (Gudger, 1931.)

It is not surprising that fairly large fresh-water fish, especially pike, should occasionally kill young, and sometimes adult, water-birds (Glegg, and Gudger, March-April, 1929), but it is remarkable that swifts and swallows should occasionally be caught by fish. (Bent, 1942.)

These birds often swoop low over water and very occasionally a leaping fish will catch one. An angler who caught a three-pound bass found, when cleaning it, that it had a whole chimney swift inside. "The digestive process hadn't begun to work, and the bird's body showed no evidence of anything that could account for his sudden end, other than a well-timed leap by the bass as the swift skimmed the surface of the water." (*Field and Stream,* October, 1939.)

Sometimes a swift or swallow will misjudge its height above the water and hit the surface (see p. 102). A fish could then capture it fairly easily. But confirmation of the suggestion that sometimes these birds are caught by fish in mid-air is found in the fact that bats, some of which skim low over the surface of pools for a drink, have been known to be captured by trout. Allen says: "The quick rush of a feeding trout is at times doubtless swift enough to intercept the bat," or, we may add, a swallow or swift. Even frogs occasionally capture birds in this way. Bent (1940 and 1942) records three such instances.

It may not be generally known that *bats* sometimes catch *fish!* The fish are caught in the interfemoral membrane or by the hook-nailed toes of the bats' hind feet, or by both, as the bats swoop over the water. (Gudger, 1945.)

Lockwood quotes the following interesting passage from a friend's letter: "We were seated by the lotus-pool when a humming-bird flew and hovered over the pool. Suddenly a bass jumped from the water and swallowed the humming-bird."

Molluscs are another under-water hazard of birds. As the birds walk in water they sometimes tread on bivalves or place their beaks between their shells. When this happens the mollusc snaps the shells together, sometimes trapping the beak or toes of the bird (plate 22).

Hopkins records finding a kingfisher with its beak held fast between the shells of an oyster. The bird's tongue was quite black through the stoppage of circulation. Cuming records a moorhen, a coot and sandpipers, and Kearton a dunlin similarly trapped. Gibbings says: "On one part of the Pamunkey River, in Virginia, it was impossible to raise ducks because at low water the ducklings got caught by the mussels and, being held, were drowned by the rising tide."

Among the records I have of birds' toes being trapped by molluscs, two are cited here. Hudson has the following account of a snipe:

> I found it on the low grassy margin of the stream with the point of its middle toe caught in one of Nature's traps for the unwary —the closed shell of a large fresh-water clam. . . . Only by

severing the point off could the bird have delivered itself, but its soft beak was useless for such a purpose. It had succeeded in dragging the clam out, and on my approach it first tried to hide itself by crouching in the grass, and then struggled to drag itself away. It was, when I picked it up, a mere bundle of feathers and had probably been lying thus captive for three or four days in constant danger of being spied by a passing carrion-hawk and killed and eaten.

The California clapper rail appears to suffer particularly from mussels. There is a mussel which lives in the mud margin just below the vegetation line on the banks of the tide channels, which is just where the rail seeks its food. When the rail steps on a mussel the shells snap together, often seizing the rail by the toe. Dawson, reporting the investigations of Chase Littlejohn, says:

> So common is this that many specimens with maimed feet or missing toes have been taken, and a few have been captured right where they were being held captive by the mussels. Others, more fortunate, in escaping, are nevertheless condemned to drag about a ball on the foot, a mass of dried mud and trash of which the mussel is the unyielding nucleus. The bivalve apparently never releases its hold, and even in death, which must soon occur, does not relax its deathly grasp upon its victim. In one instance at least, a bird was seized by the bill, and although it was able to wrest the bivalve free from its anchorage, the creature had closed upon its beak with such a grip that the bird was unable to get food, and was found in a famished and attenuated condition. This specimen Mr. Littlejohn has in his collection, a mute reminder of one knows not how many scores of similar tragedies.

I have not read of any fish being found trapped by molluscs, but small fish are sometimes caught by the lower invertebrates known as coelenterates—"hollow-bodied water-dwellers." These animals have poisonous tentacles which grapple the prey, paralyze it with their poison and then bring it to their mouths. Then it is carried into the body cavity and there digested. Young trout, cod, angler-fish and other fishes have been killed by coelenterates. Tadpoles are also occasional victims. (Gudger, 1934 and 1943.)

Another danger to medium-sized fresh-water fish during the spring comes from amorous toads. When the sexual frenzy is upon them, male toads cling with their powerful fore-limbs to the females until they eject their ova. Sometimes a male will grasp a female for several days. But if there is a shortage of female toads in a particular stretch of water, so intense is the urge to pair, that the males will seize upon anything that remotely resembles a female toad. And occasionally fish are seized in this way and, encumbered for days by the sex-maddened creature on their back, the fish sometimes die before they can free themselves. (Boulenger.)

Spiders are another occasional foe of young or small fish. This interesting fact of natural history was relatively unknown even to scientists, until Dr. E. W. Gudger collected the widely scattered references, often in obscure journals and books, to this habit and, together with information communicated to him by eye-witnesses, published a series of illustrated articles dealing with the habit (see also McKeown).

Peters has written the following account of one such attack which he witnessed:

> I saw a school of minnows playing in the sunshine near the edge of the water. All at once a spider as large as the end of my finger dropped down among them from a tree hanging over the spring. The spider seized one of the minnows near the head. The fish thus seized was about three inches long. As soon as it was seized by its captor it swam round swiftly in the water, and frequently dived to the bottom, yet the spider held on to it. Finally it came to the top, turned upon its back and died. It seemed to have been bitten

or wounded on the back of the neck near where the head joins. When the fish was dead the spider moved off with it to the shore. The limb of the tree from which the spider must have fallen was between ten and fifteen feet above the water. Its success shows that it had the judgment of a practical engineer.

Numerous other accounts will be found in Gudger's articles and McKeown's book listed in the bibliography. Sometimes the spiders have been observed to suck the juices of the fish they capture, but they do not always do this.

Spiders sometimes kill small reptiles and even mammals. In fact, some spiders appear to prefer such fare to that provided by insects. An Argentine spider (*Diapontia kochii*) specialises in the capture of tadpoles. The famous Argentine naturalist, Carlos Berg, thus describes how this is done:

> On the surface of the water, usually upon or between stones, where the tadpoles are wont to sun themselves, the spider constructs a two-winged or funnel-shaped net, a portion of which dips into the water, particularly after a rainfall, which swells the waters of the brook. The tadpoles, without suspecting the cunning of the spider, venture into the net-like wing of the tissue or its funnel, and the spider, skimming from behind upon the water, drives them on and finally overcomes one that has ventured deeper into the net. The shrivelled-up tadpole-skins surrounding the net convinced me of the skilfulness of the spider as a fisherman.

Salamanders, frogs, toads, geckos, snakes, bats and rats have all been reliably reported as being attacked by spiders. Bats, snakes and mice have been trapped and killed in spiders' webs.

While in Mexico Krumm-Heller was told of the "enmity" between rattlesnakes and a very poisonous spider. One day he witnessed the spider's attack. He writes:

A spider about five metres away attracted my attention by its peculiar behaviour. It would rapidly descend halfway to the ground from a branch to which it always returned, thus indicating that the spot where it would have touched the ground was not wholly to its liking. Looking carefully at this spot I noticed a rattlesnake, which appeared to be sound asleep. At once the stories I had heard came to my mind, and I knew what kind of spider I had before me. At last the creature seemed to have found a suitable spot on the branch, for suddenly it darted all the way down on its thread, bit the snake on the head, and then climbed as rapidly as before up to the branch. The snake for a while remained quiet, but then became very restless, writhing from side to side and rattling with all its might until it became paralyzed from the quick action of the poison injected by the spider. The rattling of this snake grew gradually weaker, its movements much slower and in less than a minute it was dead.

In the *Sunday Express* (London), January 14, 1940, there was an account from India of an elephant that was killed by bees. According to the report, the elephant had torn down the branch of a tree and had thus disturbed a hive of rock-bees, which stung the elephant in a mass. The stricken beast ran about wildly, trumpeting, and then dropped dead.

Another elephant was killed by bees in Africa. Burr, quoting one of the Reports of the Tanganyika Game Department, tells how some natives, coming across a dead elephant, cut it open and found a bees' nest lodged in its throat. Apparently the elephant, in thrusting a bundle of food into its mouth, had unwittingly picked up the nest and had been either choked, or, more likely, stung to death.

Many accidents are caused by the various agencies which may be summed up under the heading of civilization. Some of the deaths so caused, such as those of fish owing to the pollution of rivers, I shall not deal with, as wild-life mortality from these causes is too wholesale in the small area affected to be classed as accidental.

A very large number of animals are killed on the world's highways. Chalmers says he once saw a live Loch Leven trout run over and killed in the Strand in London by an omnibus! The trout was being carried across the road from an aquarium to a restaurant when it was dropped.

Beadnell estimated that before the Second World War the mortality on the roads of Great Britain amounted to an average of 10,000 animals a day, most of them birds, or over three and a half millions in a year.

In this country the respective figures were 140,000 or 50 millions, as I estimate them from figures given by Stoner. Both sets of figures are, of course, very approximate. Dreyer points out: "The rate of killing may vary greatly from year to year, and also within a single season." But the figures given do indicate, in order of magnitude, the toll of wild-life exacted by road traffic. If the unknown total of insect casualties caused by the same means are added, then the grand total for the world must reach astronomical figures.

Beadnell writes of English highways:

> Birds get slain as a rule during the day, when the traffic is at its height; on the other hand, rabbits, hedgehogs, and other nocturnal animals, suffer most at night, when traffic is at a low ebb. Pheasants and partridges may possibly be an exception, explainable by their being dazzled and frightened by head-lights, the former while roosting in near-by trees, the latter while huddled together in a field bordering the road, and then taking wing and losing their bearings, blundering on to the highway, only to get killed by the next passing car or truck.
>
> One observer describes how he counted in a single night drive within twenty-seven miles of Derby, the mangled remains of five partridges and four pheasants. On an early morning drive in 1928 I counted fifteen dead hedgehogs in four miles on a road between Aldburgh and Cromer. On another occasion, bicycling on some heights near Stonehenge, where are several warrens, I counted forty-six crushed rabbits in less than six miles. Rabbits lose their

heads, literally and metaphorically, when confronted by the fierce glare of headlights. . . .

The animals found dead on the roads were, in order of frequency: Birds, 81 per cent.; rabbits, 14 per cent.; hedgehogs, 4 per cent.; rats, 0.5 per cent.; the remaining 0.5 per cent. being contributed by frogs, toads, moles, snakes, dogs, cats, two sheep, and one forest pony.

One snake is alleged to have perished when it sank its fangs into a motorcar tire and was unable to withdraw them. The compressed air rushed into the hollow fangs, inflated the snake and blew it to pieces! (*Field and Stream,* February, 1935.)

The material used to make modern roads sometimes brings death to wild creatures. A humming-bird has been found nearly starved to death through getting its mandibles stuck together with pitch from a road. (*American Magazine,* January, 1940.) Beadnell says birds sometimes die through getting their feet "balled" by pitch from freshly tarred roads. He says he saw a thrush, that had alighted on semi-liquid pitch, held so fast that it was unable to rise.

In passing, one of the greatest, if not the greatest, death-trap for animals the world has known, was due to tar. At Rancho La Brea, in California, there are extensive tar pools in which the bones of thousands of animals have been found. The giant sloth, the bison, the saber-toothed tiger, the New World lion, bears, wolves and many other animals are represented in this great burial-ground.

It is not difficult to understand what happened. Near the rim of the pools the tar is hard, but near the center it is viscous. An animal straying onto the pools was easily trapped and would attract others, especially carnivores. They in turn would be trapped and so countless animals perished in the prehistoric past. Even today animals are sometimes trapped by the tar, and scientists in quest of fossil remains often have to free dogs and other animals from the treacherous tar. A similar tar pool has been located at Baku in the Caucasus. (Ley.)

In addition to casualties from traffic on the highways, deaths are occasionally caused by other means of transport. Trains take their toll. Kearton says a friend of his saw a covey of partridge collide with an express train and, after the train had passed, he picked up eleven birds. Kearton says, in another train *versus* bird encounter a flock of starlings hit an engine and somehow applied the air brakes, thus bringing the train to a standstill.

Birds are sometimes killed through striking the funnels and rigging of ships. Oil discharges from ships are another hazard to bird life as, if the oil gets on their plumage, it renders them flightless. Large fish and whales are sometimes rammed by ocean-going liners. Whale sharks have been reported killed by this means on several occasions. Ships' propellers probably kill numbers of smaller fish. In Chapter Three I have dealt with the casualties to birds caused by aeroplanes.

Wires of various kinds kill numbers of birds. The speed at which birds fly renders any collision with wires almost invariably fatal. It should be remembered that the wires are not so prominent to birds, viewing them from the air, as they are to people on the ground, who, of course, see them clearly against the sky.

Pangburn records finding a dead teal hanging from some telephone wires. Examination showed that the bird had evidently flown into the wires with its beak open. The open beak had scooped in two wires and they had split the corners of the teal's mouth and driven back to the base of its skull.

Twenty-nine cedar waxwings out of a flock were killed

when they collided with the one-inch mesh wire surrounding some tennis courts. (Shaw and Culbertson.) Barbed-wire fences are another hazard of bird life. Low says on three occasions he saw coots hanging from a barbed-wire fence that crossed a marsh. I have records of such fences also killing bats, insects and a flying squirrel.

Pitman says a giraffe in Northern Uganda got its neck entangled in a telegraph line. The wire became so firmly twisted round its neck that the beast was strangled. Makin says that when a 600-mile line was inaugurated in Kenya it lasted only an hour or so after the official opening. Examination revealed that one giraffe had its neck nearly severed by the line, evidently through rushing against it at speed, and three other giraffes were found entangled with the wire. The line had to be re-erected 30 feet high so that giraffes could pass underneath it!

An elk was once found with its antlers entangled in a mass of discarded telephone wire. It became so involved that it had to be shot. (*Outdoor Life*, May, 1936.) Deer occasionally get their feet caught as they endeavour to jump over fences and are held, helpless, until they die of starvation or are killed.

High-tension wires sometimes kill birds. If, for example, a flock of birds alights on some wires, their weight may be sufficient to cause two of the lines to sway together and thus cause a short-circuit, with a consequent discharge of high-powered electricity through the birds' bodies. Birds with wide wing-spreads may be electrocuted when, by stretching their wings, they make contact between two wires. Bent (1937) says many large birds have been killed in this way in this country. Allen says he saw the body of a large flying fox (an Australian bat) that had been killed in this way. Protective devices have been fitted to pylons in an endeavour to reduce the number of accidents. (Willets.)

Eagles are occasionally electrocuted in a curious way. These birds often eject their excreta in an extended stream. Should this action occur when an eagle is perched on a high-tension wire and the stream make contact with another wire below, the circuit is momentarily completed and the bird is electrocuted. Sometimes a parent bird will alight on one wire to feed one of her young perched on another. The moment the food touches the young's beak both birds are electrocuted. (Beadnell.) Birds have also been killed by high frequency radio when they have alighted on the transmitting antennae of radio stations. (Rand.)

Time, for August 17, 1942, reported a curious fatality to a horse. It nibbled at a dangling electric light bulb and was electrocuted.

Buildings of various kinds take a heavy toll of bird life, especially during the migration seasons. The mortality caused by lighthouses is well known. Their lights appear to have a powerful fascination for night-travelling birds; but flashing or red lights appear to attract them less than the fixed white lights. Incidentally, birds appear unable to appreciate the properties of glass. Shelves and perches are now erected on many lighthouses to enable tired birds to rest.

Kearton says: "On December 10th, 1882, skylarks were striking the Bell Rock Lighthouse like hail for upwards of two hours on end, during which time thousands must have perished. Upon such occasions the keepers are obliged to close every door and window in order to prevent the pressing throng of winged travelers gaining admission and knocking over or otherwise extinguishing all the exposed lights in the place." Many migratory birds die through falling into the sea owing to sheer weariness, especially when struggling against strong head winds.

Figures have been published in *The Wilson Bulletin* for several years (e.g., September, 1936) of the annual bird mortality caused by the Washington Monument, a white stone shaft rising to a height of 555 feet. When it was first erected it was sometimes possible to pick up at its base a bushel of dead birds that had been killed in one night while migrating. Later the mortality rate dropped sharply, until in recent years the average number of birds picked up at the base annually is little more than 300. Yet, owing to unexplained circumstances, on the night of September 12, 1937, 576 birds fell within an hour and a half.

The lighted torch on the Statue of Liberty in New York Harbor is a night hazard for birds. As many as 700 birds have been killed by colliding with it in a month.

Dr. E. W. Gudger, the indefatigable fish bibliographer of the American Museum of Natural History, has, with characteristic industry, collected numerous accounts of fish which have acquired various impedimenta on their bodies during their journeys through the seas. Mackerel, bluefish, haddock, gar and needlefishes have been taken wearing rubber bands about their bodies; a swordfish and a sturgeon each had a rubber band around their snouts; and four dogfish each had rubber bands round their bodies near the gills when caught. Sometimes these bands cut deeply into the flesh, occasionally almost cutting the fish in two, but often the encircled fish, when caught, appear little the worse for their strange handicap.

Both a cod and a trout have been caught with metal bands round their bodies (they had evidently pushed their heads into tin cans, the rest of which had rusted away); and a small dace in an aquarium was found to have a shell fixed on its back.

The most remarkable example of this type of accident cited by Gudger (1937) concerns a shark which was caught in Cojima Bay, near Havana, with an automobile tire encircling its body. The story was verified by Dr. W. H. Hoffmann, of the Laboratorio Finlay. He communicated his findings to Gudger, who writes:

> The shark was captured and brought in alive by two boys. They saw it out in the bay, fighting and lashing the water, and even trying to jump above the surface in its efforts to free itself. Seeing that the fish was helpless, the boys got a noose about it and thus ignominiously brought it to shore, where it was photographed [see plate 29]. . . . The tire was only removed by cutting the shark to pieces.

Gudger gives the following explanation of how he considers this strange accident occurred.

> It seems that, in Havana as in New York, street sweepings, garbage and all sorts of junk are loaded into scows, which are towed out to sea and there dumped. Sharks are scavengers, and they swarm at these dumping-grounds to feed on the garbage. When a scowload of garbage was unloaded, we can see that our shark would dash forward and would nose about in it, seeking what it could find to check its gnawing hunger. Just in front of the floating automobile tire it smelled something edible. Clamping its big pectoral fins close to its sides to get them out of the way, I can see it as it drove straight ahead through the tire until its stiff, unbendable dorsal fin fetched up against this. Fanning hard with its now free pectorals, it pushed forward hard to free itself from this strange encumbrance, which would not let go.
>
> Then came blind panic and it fought tire and sea, as it was doing when the boys found it. Presuming that at least a day or two intervened between catastrophe and capture, we can understand that the shark, fighting the tire almost continually, unable to progress through the water and hence unable to procure food, would gradually grow weaker and weaker. And in the end, too weak to

offer much resistance, it would suffer itself to be lassoed and towed to shore by the two boys.

In his 1929 article, Gudger deals with the question of how other fish have become encircled with rubber bands. He dismisses the suggestion that they are all the result of practical jokers and concludes, "that the fish, when young, swam into, or thrust its head into, the rubber band. This came to rest around the greatest circumference of the body, i.e. just behind or over the pectoral and pelvic fins, more often, just in front of the spinous dorsal. Then, as the fish grew, the band became taut, and by reason of its resiliency cut a groove in the skin and flesh."

The rubber bands themselves probably come from sewers, and the garbage and street-sweepings of New York, which are loaded on to scows and towed some twenty miles out to sea, where their contents are dumped overboard. Many of the rubber bands would float on the surface, where they would be encountered by surface-feeding fish, such as mackerel, which appear to be the chief victims of these strange accidents.

One of Gudger's articles (1938) is entitled *The Fish in the Iron Mask*. It concerns a fifteen-pound cod that was seen with its head in a syrup tin, wriggling about, trying to push the tin along on the bottom, in six feet of water at a ferry landing on Puget Sound, Washington. The fish was easily secured with a fish spear by A. B. Burnham and then carefully examined.

The cod's head was tightly held by the edges of the top of the tin where this had been cut to get at the contents. Apparently, while seeking food, the fish had thrust its head into this opening and the irregular jagged edges had prevented it from withdrawing. The underside of the tin was worn thin and in places right through, evidently as a result of the actions of the cod pushing it along the sandy bottom. The metal had worn the flesh away to the bone in a half-inch wide strip all round the cod's head.

These facts would appear to indicate that the cod's head had been imprisoned for a considerable time, yet when caught it was apparently in good condition and well-fed. The cod had, therefore, been able to feed, probably on small crustaceans which might have entered the tin seeking food themselves. As supporting evidence, Gudger says there have been a few authentic instances of mouthless fish which have managed to feed themselves tolerably well. They apparently secured their food through their gill-openings.

Incidentally, a hen has been known to live for several weeks and even gain in weight, with its head cut off! An axe had cut off most of the skull, but had left intact one ear, the jugular vein and the base of the brain, which controls motor functions. This decapitated hen walked, flapped its wings, preened its feathers and even tried to crow, but succeeded in making only a few croaky gurgles.

The hen was fed with milk and water from an eye-dropper through its esophagus, which had not closed over. Small grains of corn were also fed to it occasionally. The story of this remarkable hen, together with several photographs, appeared in *Life* for October 22, 1945.

I do not know whether Puget Sound has a particular attraction for strangely afflicted cod, but *Life* for July 21, 1941, published a photograph of a cod which was caught in the Sound wearing a pair of horn-rimmed glasses! The glasses were hooked over the cod's nose and the earpieces were caught in the gills. The glasses were later identified by their owner as a pair he had lost while fishing.

But fish are not the only creatures to become embarrassed with some of the appurtenances of civilization. An alligator, nearly six feet long, was found dead with its front legs closely pinioned to its body by a pail. The alligator had apparently thrust its head through the pail's rusty bottom while swimming. (*Nature Magazine,* August, 1926.)

A household cat, that had been missing for several days, was subsequently found with its head stuck fast in a jam jar. The cat's head was so firmly fixed that the jar had to be broken to release it. Inside the jar was a mouse—dead. Had help not arrived, the cat would undoubtedly have died. (*The Field,* April 1, 1944.) A mouse has been found dead with its head wedged in a mutton bone. It had probably been seeking the marrow and, in its eagerness, had forced in its head until it had gone too far to withdraw. (*The Field,* January 6, 1934.)

A photograph was published in *Life,* for January 1, 1943, of a cow with its horns sticking through an apple barrel. Evidently, in its eagerness for apples, it had pushed its horns through the thin wood of the barrel. The next morning the cow was found wandering in circles half a mile from its home, with agitated *moos* reverberating from the barrel. The same magazine published in the issue for February 15, 1943, a somewhat similar photograph of a deer with a bucket wedged firmly between its antlers.

One of the strangest examples of this type of accident I have heard of is recorded by the well-known Game Warden of Uganda, Capt. C. R. S. Pitman, who says that the story was supported by photographic evidence. The incident concerns an emaciated, mangy lion, which, while prowling round the outbuildings of a Government Station in Northern Rhodesia, got its head through a lavatory seat and could not get it out again! The lion was shot with the seat firmly fixed round its neck like a horse-collar.

A small number of animals lose their lives each year through colliding with the balls used in various sports. By far the greatest number of fatalities of this sort occur in golf. In fact, if one each of all the species killed by golf balls— they range from small birds to a cow—were stuffed, they would form a fair-sized museum. A reference to *The Golfer's Handbook* provides interesting reading in this connection, including details of the two-pound trout that died through jumping at the same moment as a speeding golf-ball flew over its stream! Birds, and occasionally bats, are hooked by fly- and other fishermen at times, testimony to which can be found in occasional letters to various wild-life journals.

Lincoln (1931) tabulated the causes of death, as far as they were known, of 1,290 banded birds in this country. The numbers of such deaths due to causes which may be classed as accidents, as I have used the word in this chapter, are as follows:

Flying into windows, wires, etc. - - - - - -	67
Automobiles, trains, etc. - - - - - - - -	55
Drowning - - - - - - - - - - -	19
Entanglements (strings), etc. - - - - - - -	11
Miscellaneous - - - - - - - - - - -	100

Among the miscellaneous causes of death Lincoln cites the following: a grackle asphyxiated during the fumigation of a corn crib; a bluebird overcome by the fumes coming from a chimney while it was perched on the coping; a blackbird killed in a seine; two birds killed by lawn mowers; a robin killed by over-indulgence in China berries;

a California gull killed while in flight by being hit by a golf ball; and a sparrow caught and killed by a patient in a lunatic asylum.

A few further examples of such miscellaneous hazards may fitly conclude this chapter.

Kearton says he found a dead puffin that had been smothered while it was attempting to excavate a breeding-burrow in loose, soft sand. Whole colonies of bank swallows have been smothered through slides in gravel pits where they have excavated their nesting-holes. (Bent, 1942.) Chimney swifts have similarly died when they have dislodged soot. (Burroughs.)

Land birds, which normally never enter water, occasionally alight on water, or even dash themselves into it and are drowned. Such strange behavior is apparently due to the fact, discovered by pilots, that when water is perfectly calm, it is impossible to judge, at any flying level below 300 feet, how high one is flying above it. (Jokl.) Walpole-Bond says when a large pond was drained, leaving only thick black mud, swallows still swooped over it to sip water, and hundreds died through getting their wings fast glued to the tenacious slime. Other birds have died through wading into mud and getting stuck. (Dewar.)

A strange death sometimes overtakes rodents. Their incisors grow from persistent pulps, that is these teeth continue to grow as long as the animal is alive. Normally the lower incisors work against the upper ones and this action, plus the constant gnawing, offsets the continuous growth, so that the teeth remain about the same length. But very occasionally a malformation of the jaw prevents the two sets of incisors from meeting and then the teeth may grow to such an inordinate length that, after a time, the unfortunate owner of the outsize teeth cannot eat at all. Mills, writing of the beaver, says:

> These teeth grow with surprising rapidity. If accident befalls them, so that the upper and the lower fail to bear and wear, they will grow by each other and in a short time become of an uncanny length. I have found several dead beaver who had apparently died of starvation: their teeth overlapped with jaws wide open and thus prevented them procuring food.

He mentions a beaver tooth he had seen which was crescent-shaped and more than six inches long. Plate 32 pictures a rabbit skull in the National Museum of Wales, in which the incisors have similarly by-passed each other and grown to an extraordinary length.

Shadle (1936) says in a normal domestic rabbit approximately four inches in each of the two upper, and five inches in each of the two lower incisor teeth are extruded and worn away each year.

> [*The field covered by the title is very wide and I have collected more references (over three hundred) to this subject than to any other in this book. Some of the material has therefore been discarded, as to have dealt fully with it would have entailed writing a chapter too long for the scheme of this book.*
> *I have omitted all references to deaths caused by such natural phenomena as weather, earthquakes, volcanoes, etc. Not all the accidents mentioned in this chapter were fatal, but they are cited as examples of hazards which can cause death.*]

5

Strange Uses for Animals

WHEN the warriors of some South Sea Island tribes go into battle they sometimes wear a dead fish on their heads. It is worn as a helmet and is made from the dried skin of the porcupine fish.

These fish are covered with a forest of finely pointed spines, which normally lie flat on the skin. When the fish is alarmed, it rapidly inflates its body with air or water and, as it becomes distended, the spines stand out, making the fish look like a prickly balloon. The fish's abdomen and skin are capable of such distension that the fish can more than double its normal volume. In this condition it is practically immune from all save its human foes.

Should a warrior want a helmet, he captures a porcupine fish, irritates it and, when it is fully distended, kills it and leaves it to dry. The head is then cut off, the internal organs are removed, and a pad of soft material is inserted to wear next to the head. Lappets are sometimes added to protect the ears. Another fish helmet is then ready to go to war.

Gudger (1919 and 1930) says that the stiff dried skin, with its supporting and strengthening horny spines, would be quite capable of deadening, if not of warding off altogether, any blows likely to be inflicted by the weapons used in tribal wars. Stevenson says the helmets are also used as offensive weapons in butting. The skin is quite translucent, and in Japan the fish are used as lanterns.

The porcupine fish is not the only fish to be represented in the armory of the South Sea warriors. Some of them use the horny skin of the gar as a breastplate. Stevenson says this would be capable of turning a blow from a knife, spear or hatchet.

Stomachers of shark- or ray-skin are also worn. Gauntlets covered with sharks' teeth are used as both a protective device for the hands and also as a home-made knuckle-duster to strike a foe in the face. Shark's teeth are also employed in the purely offensive weapons. Spears, swords, daggers and staffs are all edged with such teeth.

Civilized man also has used shark products in the manufacture of swords. The skin of some species of sharks has been used as sword grips, the rough texture of the skin enabling the sword to be held firmly, even when the hand is covered with blood. At the end of the First World War, Germany offered for sale enough shark skin to cover 30,000 sword handles.

Shark skin has been used by carpenters in place of sandpaper; rubbed one way the keel on the denticle acts like a rather rough sandpaper, and in reverse it smooths with microscopic fineness. The durability of the skin is remarkable. Stevenson says some skins have outworn many sheets of sandpaper of equal area. Unlike other fish, which have scales, the outward covering of sharks consists of innumerable minute teeth. The inhabitants of the Amazon Valley use the tongue of the great pirarucu, which is covered with crowded, rasp-like teeth, as a grater for coconuts, etc. (Gudger, 1943.) I am told that in the West Indies the harsh skins of "old wives" fish are sometimes used for pot scourers.

Capt. William E. Young, a professional shark fisherman, gives an example of the efficiency of shark skin as a polishing agent. Several sharks were held captive, and in their struggles they threshed and rubbed against the boat. "Sides and bottom, we found later," says Young, "were literally sandpapered to a high polish by the shark-skin contact, barnacles, moss and paint being removed as clean as the original wood."

Perhaps the most surprising use to which shark skin has been put is to make a pickpocket-proof wallet. It was made with the little spines of denticle pointing upward. The wallet would, therefore, slip easily into the pocket, but the moment any attempt was made to withdraw it the denticle gripped the cloth and held firmly. In fact, the only way in which the owner could remove the wallet was by carefully sliding his fingers down its sides to prevent it from touching the cloth at all as he carefully drew it out. (Young.)

Men who travel through the Central American forests are often attacked by insects and other small creatures. One of the most distressing of these is a minute tick which sometimes burrows into the legs. Should a native become the victim of such an attack, he relies on a fish to extract the ticks!

A white traveller was with some natives when he was thus attacked. The natives advised him to lie down in the water. Although skeptical of their advice, he did so, choosing the shallowest part of a stream. What followed is best described in his own words:

> In a few moments I felt something very sharp strike against several parts of my body; cautiously raising my head and looking towards my legs, I saw a swarm of very small fish, wriggling and swimming around me, continually bobbing their little heads at my person and . . . the fish not only picked off the ticks which were outside the flesh, but actually extracted those which had burrowed beneath the skin. (*American Naturalist*, vol. 6, p. 48.)

The accuracy of this account has been questioned, but Sanderson relates a similar experience he had while in Honduras. He and a companion became the victims "of a multitude of viciously infesting ticks." Sanderson says there is only one sure way in which to get rid of them. "You undress, leave your clothes, jump into the nearest billum pool, and then try to lie quiet. The little billums get to work, and in no time divest you of every tick. The operation drives you nearly crazy, but it is better than the torture inflicted by the ticks should they get at your skin." *Billum* is a general term in British Honduras for little fresh-water fish, but Sanderson thinks the tick-freeing fish were cichlids.

It is of interest to note that chub have been observed to perform a similar function for cattle, ridding them of infesting insect parasites. (Moodie.)

Two other unusual uses for fish are reflected in their popular names. The castor oil fish of the Pacific is so named because its oil is an effective purgative and is so used by natives of some of the Pacific islands. (Gudger, January, 1925.)

The candlefish of the North Pacific is literally used as a candle by the North American Indians. The entire body of the candlefish is permeated with a peculiar fat which will burn steadily. The Indians thread a wick through the fish's body, and when this is lighted it serves as a candle. Sometimes candlefish are used as torches, and then the whole fish is set alight and burns until it is consumed. (Swan.)

A very different use for fish has been found by fashionable ladies. Enclosed in tiny glass bowls, complete with water, very small fish have been worn as jewellery. Parisiennes wore goldfish ear-rings, in blown glass bowls, during the reign of Napoleon III. In our own day the Hungarian actress, Margot Aknay, has worn exotic fish in flat glass, water-filled containers dangling from her necklace. The colours of the fish were said to match her gown! (I once consulted a medical specialist who had a miniature aquarium containing exotic fish as a wall decoration in his consulting-room.)

Insects also have frequently been used as jewellery— both alive and dead. Malayans capture some of the extremely beautiful butterflies found in their country and then tether them to their hair as ornaments. North American Indians sometimes make a necklace by threading on a string numbers of small brightly-coloured beetles. South and Central American Indians use beetle-wing necklaces, arm bands and ear ornaments. Modern jewellers also, of course, make use of parts of insects. Incidentally, butterflies are used by artists, milliners and designers to provide ideas and inspiration for new and striking patterns.

Luminous insects are perhaps the most striking examples of living jewellery. Sometimes they are confined in gauze and then tied into the hair. Another method, used by the belles of Costa Rica, is to secure the insects with tiny chains or cords and then fasten the other end of the chain into their hair, or on their clothing, with a pin. As the insects crawl about and flash their vari-coloured lights, the effect is very beautiful. (Holder, and Verrill.)

The light of these insects has been put to several practical uses. West Indian natives fasten them to their feet (and, according to some writers, also to their hands) when travelling at night through the forest, to light them on their way. Holder quotes Jaeger as saying, referring to fire-flies (Elaters):

> I feel particularly grateful to these little insects, because, during my excursions in St. Domingo, they were frequently the means of saving my life. Often has dark night surrounded me in the midst

of a desert, forest, or on the mountains, when the little animals were my only guide; and by their welcome light I have discovered a path for my horse, which has led me safely on my journey.

Verrill says that on his expeditions in South and Central America, and the West Indies, he always kept two or three light-giving beetles in a small bottle to serve as a flashlight at night. The light of one beetle, shining at full strength, enabled a newspaper to be read easily. During the Spanish-American war an operation is reported to have been carried out by the light from a bottle of fire-flies. (*Scientific American,* July, 1935.)

Strange as it sounds, some species of beetles have occasionally helped in the preparation of museum exhibits. When skeletons are first received in the museums from the field they frequently have dried flesh adhering to them. To remove this by hand from the intricate and often fragile network of bones is difficult. Museum authorities therefore sometimes place the skeletons to be cleaned in a tray full of hungry beetles. The microscopic jaws of these insects soon pick the bones perfectly clean. At one time the American Museum of Natural History, in New York, had 5,000 beetles working in this way. (*Popular Science Monthly,* May, 1940, and *Science Observer,* March, 1946.)

But many years before scientists had thus pressed beetles into service North American Indians had employed them in making wooden pipes. A suitable stick would be selected from a tree and a small part of the pith would be removed from one end. A beetle larva would then be imprisoned in the hollow thus made. As beetle larvae are great borers, the imprisoned insect slowly ate away the pith and, when it crawled out at the other end, another pipe would be ready for use.

Beetles have also been used in primitive surgery to stitch

wounds together. Salvator Furnari (quoted by Gudger, June, 1925), writing in 1845, says, that for this purpose, North African native doctors obtain specimens of a beetle which Furnari calls *Scarites pyracmon*. These insects have mandibles which terminate in two little pincers. When placed against the edges of a wound these pincers clamp together and thus knit the wound.

If such a record stood by itself, skepticism might well be justified, but there are a number of references to a similar use of the mandibles of ants. Beebe says the Guiana Indians use the giant Atta ants to stitch together extensive wounds. The edges of the wound are drawn together and the ants' jaws are applied to them. The ants take hold, their bodies are adroitly nipped off, and the jaws, looking like a row of miniature surgical clips, remain until the wound is healed.

Beebe says these ants have a mechanical, vice-like grip quite independent of life or death. A year after he had been exploring in the homes of these ants he unpacked some boots he had then worn, and found the heads and jaws of two Atta ants still firmly attached to the leather.

Baudouin (quoted by Gudger, June, 1925) gives the following account of how ants were used in Smyrna by barbers who also practised as amateur surgeons:

> The barber presses together the lips of the wound with his left hand, and applies each ant by means of forceps held in his right hand. The mandibles of the ant, being wide open and the animal in a defensive attitude, when the insect is slowly brought to the wound it seizes the outstanding surfaces as soon as it has been brought to them, sinks its mandibles into the flesh on both sides of the wound, and remains in this position, closing each mandible against the other vigorously, and consequently holding the two edges tightly to each other. Then the barber separates the head from the thorax by a snip of the scissors, and the head with its mandibles remains in place, continuing its office though the body

has fallen to the ground. The same operation is continued with other ants until there are some ten pairs of mandibles placed at regular intervals, and so the skin is stitched together by this ingenious process throughout its whole length.

Baudouin adds that the mandibles remain for three days and by then the lips of the wound are united and the heads of the ants are removed. Stitching wounds together with insects' jaws appears to have been a regular practice among the physicians of Southern France and Northern Italy several centuries ago.

The Balinese, who are great devotees of cock-fighting, use ants to clip together bad wounds received by their cocks. According to Knight's account, the procedure is much the same as in the instances given above. This form of ant-surgery is applied to other Balinese animals as well. Another use for ants, as de-lousers, is referred to on page 172.

Another surgical use for insects is the employment of blow-fly larvae (*Lucilia sericata*) to help to heal bad wounds. (*Scientific Monthly* (1932), vol. 34, p. 531.)

There can be few stranger uses for any animal than the use made of spiders by Australasian natives, for they employ some of the arachnids found in their region to make fishing-nets! Spider silk is remarkably strong. Experiments have shown that a thread 0.01 of a centimetre in diameter will support a weight of 80 grams. The silk also has elastic properties and can be stretched by nearly a quarter of its original length without breaking. Both strength and elasticity are, of course, indispensable qualities in a web designed to trap and hold the spider's prey (see p. 84).

Some of the webs found in Australasia are among the largest in the world. Ratcliffe says he once blundered into a web which spanned nearly six feet "and almost literally bounced off." He says the silk of which the web was made was almost as thick as darning wool.

A. E. Pratt, the natural history collector, who spent two years among the natives in the vicinity of Yule Bay, in New Guinea, has the following description of how the spider web fishing-nets are made and used:

> In the forest huge spiders' webs, six feet in diameter, abounded. These are woven in a large mesh, varying from one inch square at the outside of the web to about one-eighth inch at the centre. The web was most substantial, and had got resisting power, a fact of which the natives were not slow to avail themselves, for they have pressed into the service of man this spider [*Nephila?*], which is about the size of a small hazel-nut, with hairy, dark-brown legs, spreading to about two inches. This diligent creature they have beguiled into weaving their fishing-net. At the place where the webs are thickest they set up long bamboos, bent over into a loop at the end. In a very short time the spider weaves a web on this most convenient frame, and the Papuan has his fishing-net ready to his hand.
>
> He goes down to the stream and uses it with great dexterity to catch fish of about one pound weight, neither the water nor the fish sufficing to break the mesh. The usual practice is to stand on a rock in a backwater where there is an eddy. There they watch for a fish, and then dexterously dip it up and throw it on to the bank. Several men would set up bamboos so as to have nets ready all together, and would then arrange little fishing parties. It seemed to me that the substance of the web resisted water as readily as a duck's back.

An expert on spiders considers that a net made in such a manner would hold fish of only a very small size. He considers a different technique is used in making the nets which are alleged to hold fish weighing several pounds.

When Pratt's account was published in 1906 it was regarded with incredulity by many people; but today Pratt has been generally vindicated. (See, for example, Gudger, 1924, and McKeown, 1936.) In a letter to Gudger, Capt.

C. A. W. Monckton, who was a Resident Magistrate in New Guinea, said he had seen fish weighing three, or possibly four pounds secured in spider web nets, although the fish had none of the qualities of a fighting trout. Monckton said the nets have also been used for catching butterflies, moths, birds and bats.

These spider web nets appear to be invisible in water. This characteristic has been made use of by some of the natives of the Fiji and Solomon Islands to evolve a unique method of fishing. It is thus described by Guppy, writing in 1887:

> The following ingenious snare was employed on one occasion by my natives. . . . They first bent a pliant switch into an oval hoop about a foot in length, over which they spread a covering of stout spider-web which was found in a wood hard by. Having placed the hoop on the surface of the water, buoying it up with two light sticks, they shook over it a portion of a nest of ants, which formed a large kind of tumour on the trunk of a neighbouring tree, thus covering the web with a number of struggling young insects. This snare was allowed to float down the stream, when the little fish, which were between two and three inches long, commenced jumping up at the white bodies of the ants from underneath the hoop, apparently not seeing the intervening web on which they lay, as it appeared nearly transparent in the water. In a short time one of the small fish succeeded in getting its snout and gills entangled in the web, when a native at once waded in, and placing his hand under the entangled fish secured the prize. With two or three of these web hoops we caught nine or ten of these little fish in a quarter of an hour.

Deane, writing in 1921, describes a very similar proceeding which he witnessed in Fiji. Other uses for spider web in fishing are described by McKeown (1936).

Occasionally spider silk is put to a much more sinister use. It is used to make a "smothering cap." This is made as follows. A large cone of wood or cane is passed backwards and forwards through web after web of the *Nephila* spiders until the silk is thickly felted upon it in a dense and durable material. When the silk is judged to be sufficiently thick, the supporting cone is slipped out. An elongated conical cap of thickly-matted spiders' silk is thus left.

McKeown (1936) says: "The cap is used, so far as is known, for the purpose of putting adulteresses, married or single, to death. It is quickly slipped down over the head of the unsuspecting woman while she is busied about her household duties, or while sitting by the fire at night, pulled down below her chin, and tied securely round her neck. Death by suffocation soon ensues—a barbarous, but none the less effective method."

Although there is an extensive literature on bees, I have found recorded only a few unusual uses for these insects. And these all relate to fighting or crime—the last activities one would associate with these "little almsmen of spring bowers."

According to eighteenth-century writers, Daniel Wildman, The Bee Man, used bees as a defensive weapon. He was once attacked by three mastiffs when so armed. As the dogs rushed towards him he sent forth his bees, which so stung the mastiffs that they fled. (Teale.)

Bees have several times been employed in war. When Henry I was besieged by the Duke of Lorraine, the garrison's commanding general, Immo, ordered beehives to be thrown among the horses of the attacking forces. The bees caused such confusion that the siege was lifted. Richard Coeur de Lion reduced the citadel of Acre by the same means. He ordered hundreds of beehives to be hurled among the defending garrison. To escape the stings of the mishandled, and therefore enraged bees, the defending soldiers fled for refuge to their cellars. Before they emerged, Richard's followers had battered their way into the defenseless citadel.

Even in modern times bees have put soldiers to flight. In India a cavalry man put his lance through a beehive and the disturbed insects swarmed out to the attack. Their nearest objective was a company of Highlanders, whose bare knees were easy prey. The soldiers broke and ran! I have read that during the First World War, German troops in East Africa used bees to fight our soldiers, but how this was done was not stated. Incidentally, in labor disputes in this country, bees have been used to discomfort guards and picket lines.

The crime with which bees have been associated is smuggling. An instance of this occurred during the Second World War, after the collapse of Italy. A Swiss trader and bee-keeper wanted to smuggle into his country some Italian honey, but the difficulty was how to get it across the frontier. He decided to make use of a well-known trait of bees.

He managed to get an uncensored message through to his Italian honey merchant, in which he asked him to bring all his honey-pots to the edge of a forest near the frontier and leave them there uncovered. The trader then moved his bee-hives to a point near the frontier opposite that part of the forest where the honey-pots had been put out. About 1,000 yards separated the bees from the honey.

As described on page 218, when bees find a rich, ready-made source of honey they concentrate all their energies upon exploiting it to the full. That is what happened in this instance. Within three days the Swiss bees smuggled over 200 pounds of Italian honey across the frontier under the very eyes of the unsuspecting frontier guards. (*The Times,* London, October 18, 1945.)

Camels also have been used for smuggling. Their Arab owners obtained a number of heavy zinc cylinders, each about six inches long and one and a half inches in diameter. These were filled with hashish and opium, and then forced down the camels' throats. The cylinders were weighted inside with lead to prevent them from being regurgitated.

The cylinders passed into the camels' first stomach, which possesses little or none of the digestive functions of the other stomachs. Here the cylinders could lie for weeks without serious inconvenience to the camels. The camels were then taken past the Customs officials and marched to Egyptian meat merchants who bought and sold them. When the carcasses were cut up, the tins of narcotics were easily extracted from the camels' stomachs.

But the Egyptian Customs officials received information of this ingenious attempt to smuggle narcotics and they seized a number of camels which had aroused their suspicion. On cutting one open, twenty-seven cylinders were found in its stomach. Subsequently, a proportion of each batch of camels coming to market was X-rayed to see if there were any suspicious-looking objects in their stomachs. (*Annual Report for the Year* 1939—*Central Narcotics Intelligence Bureau,* Egyptian Government.)

The Director of the Bureau has advised me that X-rays have not proved satisfactory for this purpose, as the thickness of the camel's body prevents the penetration of the most powerful rays available in Egypt. An instrument called the "Metal Detector" is now in use, which gives a special sound when any piece of metal is brought within a given distance of it.

The Director writes:

We recently [1942] experimented with this machine: we paraded three camels that Dr. Shukri, of the Veterinary Department, had previously prepared, one with nothing inside, one with one cylinder, and one with four. The experiment was a complete success, and the instrument, while keeping silent as the empty camel passed, gave forth the high-pitched note as the one tin and four-tin camels passed.

Another animal accomplice in crime is the Indian monitor or lizard. Large specimens of the genus are sometimes used by native burglars as living grapnels. The burglar ties a rope round the monitor's loins and sends it up a wall in which there is a convenient crevice. The monitor enters this and holds strongly enough to support the weight of the burglar, who then scales the rope. A similar use of this animal has been made in war. Sterndale says an otherwise inaccessible Mohammedan fort was once taken by Mahrattas using monitors in this way. (See also Kipling.)

Animals have, at times, played an even more active part in crime. Two such instances were reported in *Animal and Zoo Magazine* for October, 1938. The first concerned a magpie. It is well-known that these birds often pick up and hoard any objects that glitter. Taking advantage of this trait, a woman in Chicago trained a magpie to go into rooms in a hotel and bring to her flat any bright objects it found. The jewellery losses in this hotel soon mounted alarmingly, and, of course, the house detectives had not a clue to the identity of the thief.

Detection came when a lady, who was a particularly light sleeper, was having a nap in her room after lunch with the window open. She was suddenly awakened by a low noise and was surprised to see a bird flying round and round the room as if looking for something. Seeing some jewels lying on the table, it swooped down and seized a diamond ring. It then flew out of the window and the lady watched it fly into another window of a nearby flat. She informed the police, who, on raiding the flat, found a large store of jewels.

While I have no reason to doubt the substantial accuracy of this account, the following comments on it were made by one of the ornithologists in the American Museum of Natural History: "The magpie is an undoubted thief, but is not famed as a retriever. What it carries off it secretes, drops, or uses for its own purposes, but I never heard of one systematically taking such things back to its master or mistress. With the large collection said to have been found in the flat, I venture to say that some would have been dropped *en route* and found beforehand. Also, the bird would not be likely to have flown around the room looking for something to steal. It would have alighted on the stand or whatever it was and picked about among the various articles before flying off with one of them. Its unerring selection of the diamond ring, and, presumably, the other jewels, is a little difficult to believe."

The other case reported was as follows. Householders in the Italian quarter in New York were at their wits' end to account for a series of unusual burglaries. All sorts of articles disappeared from shops and houses, and apartments fifteen stories high had things mysteriously spirited away.

For two months this very agile "cat-burglar" ravaged the district. Then one day the little son of a detective saw a small figure ambling along the pavement. The boy looked closely and suddenly realised he was looking, not at a fellow human, but at an ape. Suddenly the boy saw it make a leap and clamber into the open sky-light of a shop. A few seconds later it emerged carrying round its neck a sack, which appeared to be nearly full.

The boy followed the ape and saw it disappear into a tumble-down house. He ran home and told his detective father what he had seen. His father acted quickly, with the result that not long after the owner of the ape, a man named Grandi, appeared in court. And this was the tale that was there unfolded.

Grandi said he had purchased the ape while abroad. It was a chimpanzee, perhaps the most intelligent of all animals. Grandi brought it home and it became a playmate for his children. Owing to its cleverness it was called Socrates.

The Grandi family fell on hard times. One day, when their food had all gone, Socrates went for a stroll on his own and, when he returned, he had his mouth full of bread and in his hand was a paper bag full of pastries. Grandi, driven by hunger, decided to exploit the ape's abilities. Socrates proved an apt pupil. With the special sack Grandi made for it, Socrates sallied forth and after a while duly returned home with the swag. The verdict of the court was: prison for Grandi, and the local zoo for Socrates.

Owing to the high intelligence of the monkey family, it is not surprising that man has pressed numbers of its members into his service. One of the most interesting and ancient of these services is to act as harvesters. In paintings on the tombs in the valley of the Nile, *circa* 2,000 B.C., there are illustrations of monkeys gathering figs and palm fruits for their masters.

Sir Gardner Wilkinson, writing in 1879, refers to this, as well as another strange use for monkeys, in the following passage, which I have slightly abbreviated from the original.

Monkeys appear to have been trained to assist in gathering the fruit, and the Egyptians represent them in the sculptures handing down figs from the sycamore-trees to the gardeners below. Many animals were tamed in Egypt for various purposes, and in the

Jimma country, which lies to the south of Abyssinia, monkeys are still taught several useful accomplishments. Among them is that of officiating as torch-bearers at a supper party; and seated in a row, on a raised bench, they hold the lights until the departure of the guests, and patiently await their own repast as a reward for their services. Sometimes a refractory subject fails in his accustomed duty, and the harmony of the party is for a moment disturbed, particularly if an unruly monkey throws his lighted torch into the midst of the unsuspecting guests; but the stick and privation of food is the punishment of the offender; and it is by these persuasive arguments alone that they are prevailed upon to perform their duty in so delicate an office.

The pig-tailed macaque, a highly intelligent monkey, is employed in Malaya and Sumatra to pick coconuts for the natives. Shelford, who has seen the monkeys thus employed in Borneo, describes the *modus operandi* as follows:

A cord is fastened round the monkey's waist, and it is led to a coconut palm, which it rapidly climbs. It then lays hold of a nut, and if the owner judges the nut to be ripe for plucking he shouts to the monkey, which then twists the nut round and round till the stalk is broken and lets it fall to the ground; if the monkey catches hold of an unripe nut, the owner tugs the cord and the monkey tries another. I have seen a Berok [the Malayan name for the macaque] act as a very efficient fruit-picker, although the use of the cord was dispensed with altogether, the monkey being guided by the tones and inflections of his master's voice. [Quoted by Gudger, 1923.]

Mann, who saw the monkeys at work in Sumatra, says the animals work industriously, but if they feel they are being worked too long they are sure to sulk! In one village he noticed a fine macaque, and asked the woman who owned it if she would sell it. She said definitely, "No," and added: "If I should sell you that monkey my husband would have to work"!

Botanists in Malaya have employed macaques to climb

tall trees and collect specimens. A string some 200 feet long is attached by a swivel to a collar round the monkey's neck. Instructions are given in the native language, and the monkey understands such commands as "Go up the tree," "Pull that twig," "Come down," and several other simple directions necessary for it to obtain the specimens required. Corner says one of his monkeys was able to find in the trees, flowers and fruits, specimens of which had been shown it on the ground, and it knew the meaning of eighteen words of Malay. (See also *The Annual Report of the Director of Gardens, Straits Settlements,* Singapore, 1937.)

It is said that the ancients made use of the skin of a hedgehog as a clothes brush. In modern times the spines of both hedgehogs and porcupines have been used as phonograph needles. (Vere, and Bates.) Fox mentions an unusual use for another kind of spine, that of the sea-urchin. He says the sea-urchins found off the Marquesas Islands, in the Pacific, have spines which are five or six inches long and the native children use them as slate-pencils.

Readers of medieval lore will know that a favourite ingredient of many medicines and potions of that epoch came from bats. (Allen.) But it is surprising to learn that these animals took a practical part in the Civil War. During the latter years of the struggle, the North effectively blockaded the sources for making niter, an important ingredient in gunpowder. The South found a substitute in bat guano —presumably from the free-tailed species, which is sometimes called the "guano bat." Big deposits of this now militarily important substance were found in large caves in Texas and, so valuable were they to the army, that the Confederate government detailed the larger part of a regiment to guard them. (Norris.)

An ingenious use has been found for one of the vilest

smells in Nature—that of skunk-oil. This is so powerful that, if the wind is right, it can be recognised half a mile away.

In many large mines in the west artificial skunk odour (butyl mercaptan) is used to carry fire warnings. As soon as a warning becomes necessary a few drops of the liquid are injected into the air-circulating systems. These create a vapour that is shot through the ventilation lines at thousands of feet per second, flashing a silent warning all over the mine, including places where no audible warnings could be heard. (*Popular Science Monthly,* January, 1939.)

Silk purses from sows' ears: yes, even this apparent miracle has now been performed. It was accomplished by the ingenuity of chemists of the Arthur D. Little Laboratories in Boston, Mass. A hundred pounds of sows' ears were obtained from a Chicago meat-packer and then the chemists got to work.

First they examined how the silk-worm makes silk. It was found that the worm emits a viscous liquid which, on reaching the air, turns into silk thread. The liquid was found to be much like glue. The chemists then took glue from a sow's ear and by various processes made it approximate to the glue from the silk-worm. From then on it was fairly straightforward laboratory work to make the silk to make the purse. The result, to quote one of the chemists engaged on this original piece of research was: "We made a silk purse. No, it isn't very good silk, or very strong, but it *is* silk." (*National Geographic Magazine,* July, 1936, and a booklet issued by the Little Laboratories.)

I have read of an alleged remarkable use of toads by German violinists. It is said that, when suffering from moist hands, they sometimes handle live toads to check the perspiration! I have tried to verify this story. I have not been able to obtain direct confirmation, but it is of interest to note that a paper by Hamet indicates that toad venom injected into a mammal has a marked local vasoconstrictor action on the blood vessels. (*Comptes Rendus Acad. Sci.,* Paris (1942), vol. 215, p. 448.) It seems possible, therefore, that a local application to the skin might check perspiration by cutting down the blood supplies of the sweat glands, thus reducing their activity.

Birds as chimney-sweeps: a less likely use for a bird could hardly be imagined, yet in Ireland geese have been used for this lowly purpose. When a cottage chimney is badly in need of sweeping, a goose is sometimes dropped down it and, during its fluttering descent, clears most of the soot away. Cuming says when a lady suggested to one of her tenants that such chimney-sweeping was cruel to the goose the man replied: "It might be, me lady. I'll put down two ducks instead, me lady."

Few birds can have been put to more unusual uses than the albatross. Old-time sailors often used to catch these birds when their ships were becalmed or travelling slowly. A hook would be baited with a piece of salt pork and thrown over the side. When the albatross seized the pork the hook lodged in its beak, but without penetrating it. To free itself the albatross would pull backwards and by keeping the line taut it could often be landed on deck. Once there it was a prisoner, for an albatross has little chance of taking off from the deck of a ship and gaining sufficient height to clear the deck rails.

Strange as it may sound, an albatross on shipboard soon becomes seasick. Owing to the oil from the sea food it eats, the vomit of an albatross is very oleaginous. Sometimes sailors made use of this oil to waterproof their leather sea boots.

The name "Cape Sheep" apparently derives from another use to which old-fashioned sailors sometimes put albatrosses. They would skin them and use the skins as feather rugs. The large webs on the feet were sometimes made into purses and tobacco-pouches, and some of the long hollow bones into pipe-stems. (Alexander.) Alfred Saunders, the Antarctic photographer, tells me that on South Georgia an albatross wing, cut off at the elbow, is frequently used as a hand-brush.

Another oceanic bird, the man-o'-war or frigate-bird, has been used by the Samoan, and some other Pacific islanders, as a postman. The natives took young birds and fed them until they were tame. Perches were erected for them near the beach. If, during the birds' later wanderings across the seas, they sighted such perches on an island, they often alighted to be fed. The natives had, therefore, in these frigate-birds a means of inter-island communication—erratic and uncertain though it was.

Turner says that one Sunday he was in a house of one of the native pastors on one of the atolls of the Ellice Group, when a frigate-bird arrived with a note from another island 60 miles away. "It was a foolscap octavo leaf, dated on the Friday, done up inside a light piece of reed, plugged with a bit of cloth, and attached to the wing of the bird." (See also Murphy.)

In addition to acting as messengers, the frigate-birds were also used as carriers. Pearl-shell fish-hooks were frequently transmitted in this way. The birds were also used to advise trading ships of what goods the natives had to offer. Hopkinson thus describes how the information was sent:

One of the places which the traders regularly visited happened to include a nesting site of the frigate-birds. It was by means of this that our islanders communicated with the outside world. Just before the nesting-season they caught a certain number of birds to whose legs they tied bladders, after which they let them go. The bladders contained pieces of stone, stick, shell, etc., which constituted the message. This always referred to the amount of "trade" (copra, pearl-shell, or whatever it was) ready for removal. . . .

In due course the birds repaired to their breeding-place. Here any would-be visitors caught any bladder-bearing bird he saw, and no doubt decided from the news thus brought whether it was worth his while making the journey to their place of origin.

6

Why Animals Act "Dumb"

"Instinct, when it operates in the normal course, when it fulfils the particular purpose for which that particular instinct exists, acts with admirable wisdom and perfection. But divert that instinct from its normal course; try to turn it into some other channel; endeavour to make it do something which it was not originally intended to do, and the result is a course of action which astonishes us by its utter folly."—R. W. G. HINGSTON.

THAT animals at times exhibit what, in man, would be called intelligence, no student of Nature will deny. But the opposite is also true. Wild creatures frequently exhibit a blind following of instinct, a complete lack of *nous*, and show, moreover, an utter inability to cope with the unfamiliar, which add up to surprising "dumbness."

Examples of intelligent action are frequently mentioned in popular literature, but accounts of the less intelligent behaviour of animals are seldom given the same publicity. It is the purpose of this chapter to set out some examples of such behaviour.

Admittedly words like intelligence and stupidity, or "dumbness" (used here mainly in the sense of slavish adherence to routine) should be used carefully in relation to animal behaviour. Some animals which behave very unintelligently in many situations, show themselves highly adaptable in others. As Dr. T. C. Schneirla comments "Most of these 'dumb' activities work wonderfully well in the animal's optimal habitat." These facts should be borne in mind when reading the accounts which follow.

The great French naturalist, Jean Henri Fabre, once took some processionary caterpillars, whose habit it is to follow one another head to tail in one snake-like string, and set them marching round the edge of a large vase. Eventually a continuous circle was formed, the first caterpillar following on the rear of the last.

Food was placed temptingly near this closed circle of marching caterpillars, but they remained on the rim of the vase for a week! With intervals for rest, they marched continuously round the vase, and completed the circuit 335 times. It was not until the eighth day that some of the caterpillars broke from the circle and then, at sunset on the same day, the procession was back home.

A similar experiment has been carried out with insects which are generally regarded as much more intelligent—ants. Wheeler confined a colony of hunting ants in a glass jar. Their habit is to go out in a long raiding column, capturing whatever lies in their path.

In the jar the same instinct asserted itself. But in this case the front of the column got linked on to the end, and thus a closed circle was formed. For nearly two days and nights the ants followed one another round the jar. Never a glimmer of intelligent thought interrupted the sequence of their fruitless march. This is how they marched to war in the open; this is how they would deploy in a circular glass jar.

Ants which got into a tight circle were observed to "mill" for nearly twenty-four hours at high speed and then to die in their tracks. (Schneirla, 1944.)

Another instance of the limitation of the ant mind was observed by Hingston. The Indian ant *Messor barbarus* has the habit of making a rubbish dump some eight inches from its nest in the ground. On this the ants place the discarded husks from the seeds they eat.

Hingston once found one of these ants' nests in a hole in a wall. He watched to see how the ants would dispose of their refuse from this unusual site. All they need do, of course, was to bring the husks to the edge of the nest and drop them over the side. But the strength of the instinct built up over immemorial ages was too great. Each ant with a husk to dispose of climbed out of the nest on to the wall, walked down eight inches, laid the husk carefully against the wall and dropped it. This useless labour continued for the months that Hingston had the ants under observation.

Hingston carried out an experiment with some dung-beetles of the genus *Onthophagus* which revealed the bankruptcy of their mental powers when faced with an entirely novel situation. It is the habit of these beetles, when they find a pad of dung, to make vertical burrows beneath it and live at the bottom. They frequently climb up into the dung, collect some of it and take it down to the burrow.

Hingston found a pad of dung inhabited by these beetles and, waiting until they had climbed into it from their underground burrows, he placed a sheet of paper between the dung and the burrow. The paper projected all round about three-quarters of an inch.

When the beetles tried to return to their burrow they were, of course, stopped by the paper. For a whole day they scratched at it, trying to get down to their burrow. None of them attempted to make a detour round the edge of the paper. After three more days some of the beetles deserted the pad of dung, others still tried to get through the unnatural obstacle, but not one had made the simple detour round the edge of the paper.

So deeply ingrained is this up-and-down movement in the nature of the dung beetles that in any enclosed space they seem absolutely incapable of movement in another plane. The experiment has been tried of placing some of these beetles in a tube half filled with sand. If the tube is placed upright, the beetles climb up and escape, but if it is laid horizontally, they try to climb up and down the sides seeking to escape, but they never attempt to move sideways, which alone would allow them to get away.

This persistence in following instinct under completely changed conditions is further illustrated by the mason-wasp. (There are several species, but their habits are similar.) As its name implies, this wasp is one of the master-builders of the insect world. It constructs its little home on trees and so colours, shades and mottles it that when it is finished it is indistinguishable from the bark. Thus does the nest escape prying eyes.

Sometimes this wasp suffers from a strange delusion. It selects the wood in houses as the foundation for the nest. In one instance a mantelpiece was chosen. Did the wasp attempt to camouflage the nest to make it blend with the polished wood or background of wallpaper? Not for a moment. For a fortnight the wasp laboured on its ornamentation and when the decoration was finished it was a perfect work of art—a very conspicuous piece of bark growing on polished wood. (Cory.)

Another wasp, which ran short of food for its larva, bit off one end of the youngster's anatomy and offered it to the other! (Enteman.)

The digger-wasp makes a burrow, places in it an egg with food, and then seals the hole. Now suppose midway through these pre-natal activities the egg and food be removed and placed just *outside* the burrow? That makes no difference to the wasp—it goes right ahead sealing up the birth chamber! One wasp was seen actually to tread on the egg as it completed the sealing operations. (Hingston.)

In the dim recesses of whatever serves a digger-wasp for a brain the immemorial order of instinct was: make a burrow, prepare food, lay egg and seal burrow. Each operation was separate and distinct. When one was finished there was no going back to put right anything which might have gone wrong with a previous operation, although failure to do so might render all subsequent work utterly useless. Like most animals, the wasp reacts only to situations as a whole; there appears to be no understanding of the relation of the parts to each other.

Hingston, however, found that a wasp (*Eumenes*) he experimented with would on occasion deal intelligently with an abnormal situation. In one instance it repeated a part of the cell-making and provisioning routine to meet the accident of a broken and depleted cell.

In a much-quoted experiment, the German investigator, H. Volkelt, placed a fly, not on the web, but in the little nest of a web-making spider. This was something new in the experience of the spider. It fled.

Demoll's interpretation of the spider's reaction is interesting. He maintains that fly-killing by the spider follows a set and rigid pattern: first the trembling of the web causes the spider to run out to see what has happened, then comes the pouncing on the trapped fly and securing it. Only after these actions have taken place does the spider kill and devour the fly. By short-circuiting these progressive steps Volkelt presented the spider with a situation with which its strait-jacketed mind was quite unable to cope. Flight was the only solution.

In a letter, Dr. T. C. Schneirla tells me that in another study, "the difference in reaction to the fly was found to be due to an effective difference in amplitude of the vibratory effect in the web and in the pocket. This experimenter duplicated the behaviour, in whatever situation, merely by varying the intensity of vibration: if weak, the spider pounced or at least remained; if strong, the spider cleared out."

The homing instinct of bees is frequently remarked upon. These wonderful little insects have been proved by experiment to be capable of winging their way back to the hive from a distance of two miles with their eyes covered. (Bonnier.)

In view of such wonderful homing ability operating over miles it is surprising to learn that a change of *inches* in the position of the hive can throw the bees into confusion. Yet such is the conclusion we are forced to as a result of Buttel-Reepen's and Wolf's experiments. It has been found that when the entrance to the hive is raised or lowered by a foot, returning mature bees go unerringly to the exact spot where the entrance *used to be.* It takes them hours, and sometimes days, to find the new entrance. But young bees, which do not know the surrounding country well, fly direct to the new position of the hive.

It is a surprising fact that some birds can similarly mislay their nests if they are moved only a few feet from their original position. On returning, the birds look for the nest where they left it and, not finding it there, they may entirely ignore it in its new position.

A good example of such behaviour occurred with a pair of grey-breasted martins in British Guiana. When the birds were away the nest was moved a few feet and placed in a conspicuous position. When the martins returned they looked for the nest in its original position and paid no attention to it in its new site. The next day they began building a new nest on the old site. (Hartley, quoted by Beebe.)

Smith records an amusing example of soldiers who

similarly "mislaid" their Nissen hut. For many months the hut had stood on the right-hand side of a road, but one day, while the men were away, it was moved to the left side. "On returning at night [the soldiers] marched right up to the old spot where the hut had stood and insisted that *their* hut was, or should be, at that precise spot, even though it was in full view just across the road!" But, unlike the birds, they did not refuse to enter, or start to build a new home on the old site!

Howard experimented with a yellowhammer. He moved the nest, with young in it, four inches from its site. When the hen bird returned she flew past the nest, went to the site where it used to be and then flew away without feeding her young. Later she returned and fed the young.

Howard then carried his experiment a stage farther. He returned the nest to its original site, transferred the young to another nest beside it, and put four blown eggs into the original nest. When the yellowhammer next returned she ignored her young and started to brood the blown eggs! But she did later feed her young when they cried for it.

A further breakdown of avian intelligence during nestingtime is sometimes seen when the birds attempt to make their nests on such a man-made structure as a ladder hanging against a wall. To the bird all the neat little boxes formed by the rungs and sides of the ladder and the wall look alike, and the bird appears to be unable to make up its mind in which of them to make its nest. Thus Hawkins found a blackbird had built *nine* nests and laid the foundation of another between the rungs of a ladder against a stable-yard wall. Six of the nine nests were completed and three were well advanced.

Boyd found that the nine uniformly shaped ventilation-holes in the side of a farm-building similarly confused a pied wagtail. It built six complete and lined nests, but laid in only one. Examples of similar multiple nest-building have been recorded for several other species of birds. (Fisher.) Jourdain says:

> An analysis of the records shows that none of these cases have occurred when the birds have been breeding under purely natural conditions, but invariably when in contact with objects constructed by man, such as the rungs of a ladder, or the spaces between the rafters inside a shed. It is evident that the intelligence of the bird is incapable of coping with the situation, and that they fail to distinguish between sites close at hand which are exactly similar in character . . . it is the artificial nature of the sites (which exactly resemble one another owing to the fact that they are man-made) which causes confusion in the bird's mind. Such exact resemblances do not exist in nature.

The cuckoo, as is well-known, has the habit of placing its eggs in other birds' nests and thus getting its young reared by proxy. When the young cuckoo hatches, its first job is to throw its foster brothers and sisters out of the nest so that it can get all the food that is brought home. It frequently happens that when the foster mother returns from a foraging trip, her own young are lying squeaking a few inches from the nest, while the young cuckoo is its sole inhabitant.

The behaviour of the mother bird in these circumstances has been witnessed (and even filmed) too many times to leave any reasonable doubt as to her reactions. She completely ignores her own young and concentrates on feeding the young upstart in the nest.

To appreciate this behaviour fully, imagine a human mother leaving her baby with a pet gorilla in the nursery for half an hour while she goes out shopping. When she returns she sees and hears her baby yelling in the street and, on going to the nursery, finds in its place a young gorilla who has just pitched the baby out of the window. She immediately forgets all about her own baby and, taking the young ape in her arms, begins to feed it.

It is true, of course, that not all birds act as unintelligently as the foster-parents of young cuckoos, but some of their other actions, when judged by human standards, are the reverse of intelligent. Blackheaded gulls can sometimes be induced to accept stones, golf or cricket balls, or even tin receptacles in place of their eggs. (Kirkman.) Penguins have likewise been persuaded to brood chunks of ice. Levick says he once found a penguin sitting on two eggs in the middle of a small river, presumably caused by melting snow. He lifted the eggs out and put them on dry ground close by. The penguin completely ignored the eggs after this.

It is questionable if any intelligence enters into the brooding of a bird's eggs. During this period "brood spots" appear on the bird's breast—areas which are denuded of feathers and where congestion of the skin capillaries brings about uncomfortable heat. The coolness of the eggs pressing against the brood spots relieves the bird. The constant turning of the eggs, which is a characteristic action of a brooding bird, ensures that as soon as one part of the egg's surface becomes warm it is changed for another and cooler part. These actions are admirable for hatching eggs, but the extent of the intelligence involved can be gathered from the fact that if an incubating bird's breast is immersed in cold water, thus counteracting the abnormal heat, the bird will no longer be interested in brooding its eggs. (Noble, and Tucker.)

Levick has the following interesting passage on penguin behaviour:

When conditions arose which were new to their experience the penguins seemed utterly unable to grasp them. As an example of this, we had rigged a guide rope from our hut to the meteorological screen, about 50 yards away, to guide us during blizzards. This rope, which was supported by poles driven into the ground, sagged in one place till it nearly touched the ground.

At frequent intervals, penguins on their way past the hut were brought to a standstill by running their breasts into this sagged rope, and each bird as it was caught invariably went through the same ridiculous procedure. First it would push hard against the rope, then, finding this of no avail, back a few steps, walk up to it again and have another push, repeating the process several times. After this, instead of going a few feet farther along, where it could easily walk under the rope, in 90 per cent. of cases it would turn, and by a wide detour walk right round the hut the other way, evidently convinced that some unknown obstacle completely barred its passage on that side. This spectacle was a continual source of amusement to us, as it went on all day and every day for some time.

The story of the ostrich burying its head in the sand to escape detection may be fabulous, but an almost exactly similar phenomenon has been observed in the case of a moorhen. When surprised in the open it stuck its head into a mouse-hole and crouched there, perfectly still, while its whole body was plainly outlined against the bank. (Batten.)

Dumbness of another variety is occasionally exhibited by woodpeckers in storing food for the winter. They hammer out holes in trees with their pick-like beaks and then stuff in acorns and nuts. (Stones and pebbles are indiscriminately stored as well at times.) But sometimes a particularly vigorous bird, working on a thin soft-wood tree, will bore a hole right through. That makes no difference, and in goes acorn after acorn, nut after nut, to fall uselessly on the ground on the other side of the tree. A double handful of acorns has been picked up on the ground under a hole which had been thus drilled by a woodpecker. Woodpeckers have also been observed dropping bushels of acorns into a wooden shack from which there was no possibility of the birds regaining them. (Russell, and Ritter.)

Squirrels, of course, do the same. Some flying squirrels kept in captivity had the storage instinct so highly developed that they filled up their nesting-boxes so completely that they could not get in themselves! (Hatt.)

A similar failure to understand the meaning of a "bottomless hole" accounts for the energetic but useless labour of the two starlings reported by Frances Pitt. The birds found a hole in a shutter outside an unlighted loft. Without exploring further, the starlings began to collect nesting material with which to line the hole preparatory to using it for a nest.

For how long nesting material was dropped through the hole is not known, but it was at least several weeks. When the loft was eventually examined there was a pile of rubbish on the floor which would have filled a sack; enough, in fact, to have made a dozen starling nests. And the birds were still bringing material with which to line the bottomless hole!

The classical example of mammalian dumbness concerns a cow which was bereft of her calf. To console her the calf's skin was taken, stuffed with straw, and then placed alongside the cow. She commenced to lick the skin with maternal devotion. During this process a particularly vigorous lick caused part of the skin to give way and revealed the succulent straw stuffing. The maternal instinct was at once forgotten and the cow began to eat the straw, chewing it with as much relish as ever she displayed when devouring her ration in the byre.

Just try to translate *that* in terms of a human mother ravished of her babe and you get some idea of the immeasurable gap which exists between the intelligence of a human being and a cow.

Rats are generally rated fairly high in mammalian intelligence tables, yet some of their actions betray, at times, an almost incredible silliness. A tame white rat, building its nest, ran out of material. In its frantic search for objects with which to appease the nest-building fever it stumbled over its own tail. At once this member was seized and carried to the nest. Later the tail was again discovered and again carried to the nest, and so on for a dozen or so times. (Sturman-Huble and Stone.)

Young rats frequently stray from the nest, owing partly to their habit of clinging to their mother's teats and so being dragged out. The instinct to replace straying ratlets is therefore developed in parent rats. But the utter lack of reasoning displayed in the action is shown by the fact that if fifty young rats (or young kittens for that matter!) are placed outside the nest they will all be dragged in as far as space will allow. (Wiesner and Sheard.)

I referred previously to the way in which wasps appear to react only to whole situations and to have no understanding of the relation of the parts to each other. In the following experiment Buytendijk and Hage showed that a dog appears to suffer from the same limitation in reasoning powers.

The dog was put into an apparatus which had eleven doors. If it went through the right one it was rewarded. The dog soon learned which door to go through to get the reward.

Then the whole apparatus was turned round 180 degrees. As the room had windows on both sides the illumination remained the same. Yet when the dog was now placed in the apparatus it was completely lost, and could find the right door only after seeking for it a considerable time.

"The explanation must be that the dog was not trained with regard to one single characteristic, as, for example, the direction it had to choose when entering the apparatus, but was oriented to a complex of characteristics of various kinds, kinæsthetic and visual ones, the latter from inside and outside the apparatus. As the apparatus was turned round, this complex was disturbed, and the dog had to learn anew how to find its way to the right door."

No account of dumb animals would be complete without mentioning "silly" sheep.

A shepherd was once driving some 400 sheep through a gate. A boy sitting by the roadside held out his foot in front of the leading sheep and it promptly jumped over it. So did the next sheep and several of those following. Then the shepherd made the boy withdraw his foot.

But the damage was done. The following sheep also jumped when they came to the same spot, but after about a hundred sheep had similarly jumped, something went wrong. The sheep became confused and jumped several times instead of once. The jumping mania spread and soon there was a whole line of jumping sheep stretched out for half a mile along the road.

The boy who held out his foot became in after years a famous British naturalist, E. Kay Robinson. He also discovered the reason for the strange follow-my-leader jumping complex of the sheep.

The ancestors of domestic sheep were creatures of the mountains and hills. They were often pursued by wolves and frequently their way lay along mountain paths they had never traversed before. To quote Robinson:

They must go in single file because the path is usually only a narrow ledge of rock, and perhaps the leader may suddenly come to a gap, 20 feet wide, in the ledge. Without hesitation he leaps it, and, landing with all four feet together on the opposite side, races on. The next sheep could not even have seen the gap because the leader's body was in the way; but he, too, must jump a clear 20 feet exactly where the leader did and race on.

So in quick succession each of the sheep leaps the chasm, and lands with all four feet together exactly where the one in front landed; but of the wolves, some perhaps fall down the precipice before they can stop, but none achieve the leap. So, although it may seem very silly that a whole flock should go through a gap in the hedge (or jump over a boy's foot) just because one has done so, it is really the wisdom of ages which guides them, for the two or three thousand years during which sheep have been more or less domesticated are too few to obliterate their instincts.

The reason (at least when you know it) is fairly obvious in this instance, but in the one that follows it is more subtle. Sheep sometimes develop a habit of jumping fences into forbidden territory. The shepherd's remedy is to take the ram who leads the forays and tie his ears back!

Why? Because it is an ineradicable habit with sheep to point their ears forward as they jump. Baulked of the ability to do this, they are incapable of "taking off." As other sheep blindly follow the ram, it is necessary only to cure him and the whole flock will refrain from jumping fences. (Seton.)

F. W. FitzSimons, the Director of the Port Elizabeth Museum and Snake Park, says they once owned a two-headed sand-snake (*Psammophis*), each head of which possessed three inches of neck attached to the communal body. For sheer mutual self-preservation it might have been thought that the two heads would live and let live, even if the standard of collaboration did not come up to that of human Siamese twins. But no, the reverse happened.

One night one head swallowed the other, but the two

heads were separated before any vital damage was done. Thereafter the head which had been a temporary meal apparently bore an undying grudge against its fellow. And one day it took its revenge. It bit the other head repeatedly and injected large quantities of venom. The head thus violently attacked retaliated, and in this mortal scrap was ended the tale of these one and a half very dumb animals. (This is not an isolated example of such behaviour among two-headed snakes. Cunningham, who has devoted a book to such reptilian monsters, records several instances of the two heads fighting.)

7

Have Animals a Time-Sense?

H. MORTIMER BATTEN says he once knew a gypsy boy who, although he had never possessed a watch, could always keep an appointment punctually. Moreover, at any time the boy could tell to within a few minutes what hour of the day it was. When asked how he did it the boy laughed and said: "Just know."

Whatever the explanation may be, most people appear to possess a time-sense which enables them to gauge the passage of time with tolerable accuracy. The most familiar example is the faculty by which the mind wakes the body at a certain time in the morning. If, over night, the subconscious mind is impressed with the thought that one desires to awake at a certain time, almost invariably at that time one finds oneself awake.

During an experiment to test this faculty, two people were confined in a sound-proof room in which they could obtain no possible aids to marking the passage of time. One subject was released after spending eighty-six hours in the room and the other sixty-six hours. In their estimates of how long they had been in the room one was correct to within forty minutes and the other to within twenty-six minutes. (MacLeod and Roff.)

In the animal kingdom this faculty of natural time-keeping is often found to be remarkably developed. Sir William Beach Thomas (1943) has told of a shepherd who used punctually to deposit in the pen a supply of roots at a given hour in the afternoon. He found after some time, that if he came a minute or two late, all the sheep were standing, but if he came a minute or two before the usual time all the sheep were lying down.

Fred Walker, in his interesting autobiography, gives another example of animals' acute awareness of their exact feeding-time. He was at one time working on a farm in California and used to plough with a team of ten mules. He writes:

> I never needed a watch when I was ploughing with these mules. Although the farm was not equipped with sirens or any other modern device, these amazingly intelligent animals knew to the second when it was twelve o'clock—lunch-time. They would stop dead and begin to bray. It was useless trying to goad them on and unless they were unhitched and fed they would turn the plough over in their angry revolt.
>
> At six o'clock in the evening the same thing would happen. I might be in the middle of ploughing a furrow, completely forgetful of the time, when suddenly the mules would stop, begin to bray, and then turn in the direction of home. On the first occasion that they struck work I very foolishly thought I could make them finish the furrow, but the next thing I knew was that they were all cantering for home, with the plough upset and myself holding on like grim death to the wreckage. After that experience I gave them best.

Dr. A. S. Hudson relates a somewhat similar occurrence with cows, but in this instance the time-sense operated over a week. He once had charge of five cows during the summer. They grazed in a pasture a few hundred yards from the house. They were given salt every morning. This was enjoyed and it was evident that the cows began to expect it.

"After a length of time," he writes, "a curious behaviour of the cattle became conspicuous, for every Sunday morning they were found standing at the bars, the point nearest the house, with every appearance of mute expectation. At every other morning, as well as at evening, they had to be sought and brought to the bars for milking. Sometimes I would forget to take the salt with me at the stated time, when, instead of moving off to feed after my task was done, as they usually did, they remained about the spot an hour or so, as if waiting for their weekly rations of salt."

He adds that the cows could not tell when Sunday had come by any change in routine. They were completely isolated from any human activity (there was no thoroughfare near them) and they saw no one but Hudson all the time. So far as he could judge there was nothing to help the cows to distinguish Sunday from any other day of the week, except that, of course, on that day they were given salt.

Dr. Carl L. Hubbs, who read this chapter in manuscript, says critical workers on animal behaviour may consider such anecdotes as these valueless. He adds: "But experiencing such behaviour is impressive. When I was still going to high school and was left alone for a time on a ranch in California, a cat I had raised was my closest companion. It regularly met me as I returned across the fields from school (three miles away). It would leave with me every evening when I went to a neighboring farm for milk, but at all other times refused to go with me. Some sense of time of day plus other associations seems to have been involved."

Dr. Gustav Eckstein, of canary fame, tells of the remarkable punctuality of another cat. Every Monday, at 7.45 p.m., this cat left its home, walked to a hospital and jumped onto a window-sill of the nurses' dining-room, where on Monday evenings the nurses played bingo, a game played with boards and counters. On three successive Mondays Eckstein followed the cat and watched it go through this strange routine. "I thought it might be food," Eckstein says, "but there was no food. Or a congregation of cats, but there were no cats. He was there at that exact time to hear and see the people playing."

Eckstein interviewed the two ladies who owned the cat, which was called Willy. They told him: "Oh, yes, Willy knows Monday. Any other night Willy may go out at five o'clock, or six, and not return till eight or ten or midnight for his supper, but on Monday he stays in, and promptly at 7.30 eats his supper. He leaves promptly at 7.45 and we can count on him to come home at quarter of ten, when the bingo game ends." Eckstein adds: "This I did myself observe."

This cat also appeared to know 8.10 in the morning. His mistresses used to leave home promptly at this time each day. The cat, after spending a night out, liked to be home before they left. If he had time in hand he came along slowly, perhaps stopping to stretch in the sun for a little while. If, however, his night wanderings had made him a little late, he did not dawdle at all, but came hurrying along to be home promptly by 8.10.

A golden retriever was trained to act as an alarm clock. Every morning it tapped at its master's bedroom. On four successive days this morning call was timed, and on each occasion it took place at 7.22 precisely. (Thomas, 1943.)

Many similar stories are told of the domestic dog's time-sense. Schmid once owned a wolf-hound which, whenever its master went to town, went punctually to the station to meet the train by which Schmid returned. Schmid always went to the Post Office at a certain time in the afternoon. The dog appeared to know the time to the minute, for any delay in setting out caused it to become very restless.

W. H. Hudson tells of a wild teal that became domesticated and was greatly attached to its master. At about the time he was due to return from business "she would go to the open street door to wait for his return, and if he was an hour or so late she would sit there the whole time on the threshold, her beak turned citywards."

Commander C. B. Fry tells the following story of a dog he once had which "used to meet me every Saturday at ten minutes past one at the gate at the bottom of our drive. Every Saturday, as sure as the train puffed away down the line, Joe was at the gate wagging his entire body. He never went down to the gate at any other time. It is quite useless for any psychologist to tell me that he did not know the day of the week and the time of day."

But I am afraid the psychologist might argue that there were special domestic activities associated with the return of the master around one o'clock every Saturday, which gave the clue to Fry's dog that this was the day and time of the week when the master should be met. Buytendijk, discounting the idea that dogs can know the time and when a certain day of the week comes round, says: "All this can generally be traced to an extension of the area of excita-tion brought about by experience."

Evidence that Buytendijk's explanation is probably correct is provided by the behavior of a farm dog, which promptly at seven in the morning and four-thirty in the afternoon, used to go off on its own account and bring in the cows from the pasture to be milked. But when daylight saving was introduced, although at first the dog kept to the old times, it eventually fetched the cows in at the revised times. (Batten.) To me the only explanation can be that the farm routine gave the dog the clue to the time, or else why did it change at all?

Another illustration of how the time-sense can sometimes be explained by changes in the surrounding circumstances is provided by the behaviour of the sparrows
and other small birds which, before the Second World
War, frequented the Luxembourg Gardens in Paris. The
birds used to gather punctually at 9.45 in the morning to
await being fed by a kindly visitor. The birds did not wait
for his appearance: they always gathered *before* he appeared.

But now comes the most illuminating feature of this
story. When daylight saving began the birds *put forward*
their time of gathering one hour; that is, they still con-
gregated at 9.45 a.m. by the clock, although the actual
time was then only 8.45 a.m. Commenting on this, Sir
J. Arthur Thomson says:

> This seemed almost magical, yet the explanation is probably
> simple. The work of the Gardens, such as sweeping leaves and ad-
> justing seats, went on regularly day after day, and the probability
> is that the birds established an association between what was go-
> ing on and the approaching visit of their hospitable friend. When
> the change was made to "summer time" the routine of the garden
> was at precisely the same stage as before. It is difficult to prove
> such a theory without experiment, but various carefully-observed
> cases point to the view that animals may seem to tell the time
> when they are simply observing certain signs of the times which
> have come to have a profitable associative value.

While such explanations as this may solve the mystery of
the time-sense in domestic and semi-domestic animals,
they can hardly account for the same faculty when found
in wild animals. What, for example, can be the explanation
of the strange sense of timing found in the following description by Devany of the antics of the wild spruce grouse
during its courtship display?

His favourite location at such a time is between two trees standing apart some twenty or thirty feet, and with the lower branches large and horizontal. Perched on one of these branches he pitches downward, pausing midway to beat and flutter his wings, and ascend to a branch of the opposite tree. After a short interval this manœuvre is repeated and so continued by the hour, swinging back and forth from tree to tree, the time between each swing being as exact as if measured by a watch.

Evidence of another form of time-sense is found in the behaviour of some shore-birds. Curlews, feeding on the mudflats of river estuaries, are compelled at times by the rising tide to fly inland and feed there, until the receding tide leaves the flats once more exposed. The curlews have been seen, when several miles from the shore, to cease feeding, collect together and fly back to the shore at the very moment when the shallows were first exposed.

It has been suggested that scouts, stationed within sight of the sea, give notice to the birds which are feeding inland. Commenting on this suggestion, Seton Gordon says: "I would rather incline to the belief that the birds' sense of the progress of time is sufficiently accurate for them to feel instinctively when it is possible for them to return to their interrupted meal."

A similar phenomenon has been noted with the Australian reef herons. Ratcliffe writes:

My friend, C. M. Yonge, told me a remarkable thing about the reef herons. Apparently they have developed within their inadequate-looking heads something in the nature of a self-adjusting alarm clock. Among their favourite feeding-grounds are the reefs which dot the sea inside the outer barrier and at low tide offer a fine harvest of shell-fish, crustacea, and other sea creatures. At high tide they are submerged, and therefore the herons have to leave them and fly back to the nearest land, often a great distance away.

Day after day they make the oversea journey, always arriving at the reefs just when the coral begins to show above the water. . . . Even if low tide were at the same time every day, their punctual arrival at a spot thirty miles off the coast would be miraculous enough. But it isn't: there is a daily lag of some forty-five minutes. Where are the clocks which warn them to postpone their departure in accordance with the lunar rule? Here is a sixth sense if ever there was one.

This behaviour of the reef herons was thoroughly discussed in *The Victorian Naturalist* (November, 1943, to September, 1944.) Dr. H. Flecker maintained that the birds could either see the state of the tides on the reefs (although they might be thirty miles away!), or could estimate it from the state of the tide at their feet. Another suggestion was that the herons were able to appreciate the slight changes that take place on the surface of the earth, owing to the added or subtracted weight of the water along the shore-line at the tidal changes.

There was also considerable discussion on a suggestion that "radial rays at present unknown, but due to the tides, affected the birds in such a way that they knew when to leave the mainland and reach the islands at the proper time."

Possibly Flecker is right, and there is nothing more in the behaviour of curlews, reef herons and other birds with similar habits, than can be explained by ordinary avian eyesight, which is admittedly very keen. But in view of some of the other examples mentioned in this chapter of what appears to be some sort of "self-adjusting alarm clock" in animals, perhaps there is more in the less prosaic explanations of these birds' behaviour than some ornithologists are at present prepared to admit.

The almost uncanny punctuality and regularity of animals in some aspects of their lives would seem to indicate the existence of an internal clock, rather than a reliance on more or less variable external associations.

Mann says that a chacma baboon, that lived for twenty years in the United States National Zoological Park, set its own time limit for the hours it would allow itself to be on show. Every day at four oclock it would enter the small house provided for it and close the door. Its day's work as an exhibit was over.

The punctual arrival of sandgrouse at their drinking-places has been remarked by travellers. Boyd Alexander (quoted by Bannerman), writing of the pin-tail sandgrouse in the vicinity of Lake Chad, says: "The punctual habits of these birds are remarkable. Every morning and evening at the same hour, batches would fly high over our camp, suddenly to drop down with a sound like a shower of spent bullets in the water." But moonlight nights appear to upset the birds' normal routine.

The pileated tinamou, of Panama, is popularly known as the three-hour-bird. The nickname has been given because this bird is reputed to sing at three-hour intervals during the day and night. Sands says that when he was in Panama a tinamou used to sing regularly at six-thirty every evening, and it was so punctual that watches could be set by its song.

In northern Michigan, at the University Biological Station on Douglas Lake, the regularity with which the whip-poor-wills begin to call in the late evening, with little respect to the weather, has long surprised zoologists. Dr. Charles W. Creaser, who teaches at the Station in the summer, has kindly written to me as follows about these birds:

> The whip-poor-will is a very common bird and the early part of the session (late June and early July) is their nesting season. It is a common observation that they start their call about the same

time each evening, and it is not an uncommon thing for students and staff to look at their watches to see if the first one is on time or to check the watch! However, I have not made any careful study of this or set down any data. I know that it is related to the sunlight, as it shifts as the season progresses.

Lutz noticed a similar punctuality in the first song of a wren. On normal days this occurred regularly between 5.57 and 5.58 a.m. Allard observed that two other wrens began to sing at different times each morning, but the second consistently allowed the same period to elapse before it joined its song to that of its companion. (Both references from Calhoun.)

In summer the Australian flying-foxes, or fruit-bats, congregate during the day in vast numbers among densely foliaged trees. Promptly at six o'clock each evening the bats break camp, as it is called in Australia, and stream out to their distant feeding-grounds. The punctuality of this daily manœuver is such that it is alleged a watch can be set by the time of the bats' departure. (Barrett.)

Bowles says he was once visiting a friend and happened to remark that it was half-past three. His friend quickly replied: "In five minutes it will be bedtime for our Flicker."

They went outside the house and watched. In about five minutes the woodpecker appeared and hung itself against a board under the eaves, where it spent the night. The bird had been doing this with absolute regularity for some time. Exactly at three thirty-five each afternoon, even although it was broad daylight and bright, sunshiny weather, the woodpecker always called it a day and went to bed.

Starlings, too, appear to retire to roost very punctually. Soldiers at one North of England camp used to be able to tell the time by the starlings which flew home each evening across the barrack-square.

Batten gives several examples of the same sort of punctuality. During one summer he noticed a woodcock fly past his study window each evening, always at the same time. A heron likewise flew over his house each evening just before sundown.

Another example he gives concerns a hare. He writes:

> At one time I used to motor to and from the station each morning and evening, and that autumn I used each evening to see a hare on his way to his feeding-ground across the meadow. He was so punctual that I might have set my watch by him. If my train was to time he would be in the middle of the pasture—generally sitting up and listening as I passed. If my train was a little late he would have crossed the meadow, and perhaps I would be only just in time to see him pass under the gate on the other side.

It is a far cry from a Scottish country lane to an equatorial forest, yet here also the lives of numbers of animals appear to be lived to a fairly rigid time-table. Sanderson writes: "Night after night, for weeks on end, I have watched the same squirrel, genet, bats and monkeys pass the same spot, always travelling along the same branches, following almost footstep for footstep the same path. Each passed me at the same time to within a few seconds, though the weather on several occasions varied greatly."

Another aspect of the time-sense in animals, closely allied to that of punctuality, is the so-called rhythmical response. Sea-anemones, for example, open at high tide and shut at low tide. And there are marine worms which come to the surface when the tide ebbs and disappear again with the first splash of the flow.

Bohn found that when these animals were placed in tideless aquariums on shore, they reacted in exactly the same way, the anemones opening and closing, the worms appearing and disappearing at the correct time for the tides. But, as far as I know, Bohn's experiments have not been verified by later workers who have carried out similar tests.

A remarkable time-sense is exhibited in connection with the spawning of two marine animals, the palolo worm and the grunion.

The palolo is an eunicid worm found in the Pacific. There are several species, but it is in *Eunice viridis* that the time-sense is most fully developed. This worm lives in deep, cavernous hollows at the base of sunken coral reefs in the waters round Samoa, Fiji and some other Pacific islands south of the equator.

Once only each year, at a definite time, the palolo appears in myriads at the surface of the sea to perpetuate its species in a spectacular swarming. This takes place in the early spring, exactly one week after the full moon in November (springtime in the south Pacific) and occurs with such regularity each year that "palolo time" is the most outstanding date in the native calendar, important events being counted from the date of the swarming. According to Woodworth a few palolo rise during the October quarter of the moon, but their numbers are in no way comparable to the November swarming.

The worms grow to a maximum length of eighteen inches. As November approaches the hind part of each worm, which is about three times as long as the fore part, becomes filled and distended with minute eggs in the female and sperm in the male. This end part gradually increases in size until the moment for spawning.

When that moment arrives each worm crawls backwards out of its hole deep in the coral and the hind part breaks away and wriggles up to the surface. The fore part of each worm remains in the coral and grows a new hind end which, the following November, again supplies the eggs or sperm for the perpetuation of this strange species.

Almost immediately the hind end of the palolo reaches the surface, it bursts and the eggs or sperm are fired into the water. Woodworth, who observed the discharge under controlled conditions, says: "The process was like an explosion." The empty, shrunken remains of the worm then sink down to die on the sea-bed. The great majority of the countless millions of palolo worms inhabiting the coral reefs in the South Pacific behave in this way once a year, in the early morning of the seventh day after the November full moon. Burrows says the palolo "makes its annual rising at an approximate date by the calendar year, but at an actual date by the moon and tide."

The time of the annual swarming is a great occasion for the local natives, as the palolo is regarded as a delicacy. Basil Thomson has given a picturesque account of an excursion he made with some natives to gather some of the worms when they swarmed. They sailed to a deep coral reef and arrived there about half an hour before the swarming was due to begin.

Thomson looked over the side and exactly at the expected time he saw, deep down in the clear water, what looked like a thick jet of smoke streaming out of a hole far down in the coral. The jet mushroomed out as it reached the surface and the water over a wide area began to look like vermicelli soup, so alive was it with wriggling, bursting worms. Soon the sea became milk-white with the contents of the disintegrated worms. Fertilization began at once and later the cloud sank to the bottom almost as quickly as it had risen, leaving the surface of the sea as clear as it was before.

Thomson filled a glass tumbler with water containing

the reproductive segments as soon as they reached the surface and watched the whole process take place. He says fertilization in the tumbler synchronized exactly with that in the sea. The water in the tumbler became clear, with a sedimentary deposit on the bottom, exactly when the sea once again became clear.

The natives, meanwhile, were busy scooping at the surface with large grass baskets. By the time the swarming was over, each canoe was filled with as much as it would hold of palolo remains. During the return journey, cakes of this gelatinous mass were fried and, says Thomson, tasted just like oysters. Burrows also found them "very good eating," and says, if you can forget what they look like before being cooked, they are delicious.

Burrows, who attended a rising and described it as an "unforgettable event," adds some interesting information which I have not seen mentioned in other accounts of the palolo. He says:

> All round and between the boats big fish and sharks cruise quietly along, gulping them in, and take no notice whatever of the boats or their occupants. . . . A curious fact is that all fish caught in the neighbourhood of the rising are poisonous to human beings for about ten days or a fortnight after the event.

The other marine animal which exhibits a remarkable time-sense in connection with its spawning is the grunion, a species of small silverside found only off the coasts of Southern and Lower California. Beginning in March and continuing through July, the female grunion ripens a mass of eggs. This ripening is correlated to a remarkable degree with the tides, which play an important part in the spawning of these fish. Clark (1938) says: "Spawning occurs only every two weeks and the time required to mature a batch of eggs is so mysteriously adjusted that the fish are ready to spawn only on the three or four nights when occur the exceptionally high tides accompanying the full and the dark of the moon."

Dr. Carl L. Hubbs, another authority on these fish, writes to me: "The ripening rhythm may be related to the spawning on time, but can't be the whole explanation, for the cycle starts each year."

On three or four nights following each full moon and each dark of the moon, that is when the tides have just passed their highest for the two-weeks' period, the grunion come to the wave margin on the beaches to spawn. The biggest runs occur on the second and third nights after the moon is full or dark, and last about an hour.

Like the swarming of the palolo, the pairing of the grunion is a spectacular performance. Waiting until a few minutes after the highest point of the tide, the grunion ride one of the larger waves and, swimming through the surf, are washed ashore. There, at the farthest reach of the tide, on the wet sand far above the limit of the average tide, male and female grunion go through the spawning "dance."

The female digs itself vertically tail-downward into the wet sand, until only its head protrudes, and then extrudes her eggs. The male, or males, then curve their bodies alongside the female, keeping horizontal and at the surface of the sand. The sperm they eject moves down through the soft wet sand to meet the eggs, which are fertilized in high percentage. Funkhouser, who has often watched the performance, says it looks as if the two grunion are dancing on their tails. After the spawning, the male flops and wriggles away towards the water, and the female, exhausted, sways back and forth until she has freed herself from the sand

and is washed back to sea by the next wave. The whole process lasts about thirty seconds.

But, as with the palolo worm, the local inhabitants regard the grunion as a delicacy, and on the nights when there is a grunion run, as the spawning is called locally, the beaches are lined with people intent on catching grunion. Fires are built and beach parties feast on the fish. To conserve the fish, grunion may be caught only by hand: nets, buckets, seines and other devices are prohibited. The runs during the middle of the spawning season are closed to all fishing. There is no commercial fishing for grunion.

In view of the popularity of the grunion runs, and also because of their punctuality, the local newspapers publish the times of the expected runs. They cannot forecast the beaches on which spawning will occur, as these vary. The schedules read almost like a railway time-table, saying, for example, that "A grunion run is expected at about nine-twelve this evening." Occasionally the schedules go wrong by twenty-four hours, but on the whole they are remarkably accurate.

Funkhouser thus describes how these schedules are worked out:

> Predicting grunion runs almost to the minute seems something of a mystery to the uninitiated, but when one knows their habits it is simple. All one needs to know is the time and the height of tides, and many calendars have these printed on them. Since grunions spawn only at night, the day tides can be disregarded. Tide height reaches peaks at fourteen-day intervals, once at the full of the moon, and again at its dark. The forecasting of the hour and minute when grunions will run is reached by adding fifteen minutes to the time the tide reaches its nightly peak. In other words, there is a lag or margin of safety: they come ashore after the turn of the tide, and on nights when the tide reaches a little less high than on the preceding night.

Because the eggs are laid at the margin of the full moon high tide, they remain on the beach until the next high tide, two weeks later. After about ten days the eggs are ready to hatch but they are not actually hatched until the waves erode the beach and uncover them. As soon as they are free in the water, they hatch explosively.

Should the eggs not be washed out by one tide, they can remain buried in the sand without harm for another fortnight, until the next series of tides completes the uncovering process, whereupon the eggs hatch with the same explosive suddenness.

A few workers have undertaken laboratory investigations into the time-sense in animals. In his oft-quoted experiments on the conditioned reflex in dogs, Pavlov found that by ringing a bell every time food was offered to a dog, the dog's mouth could eventually be made to water merely by ringing the bell, even though no food was offered. In further experiments Pavlov trained a dog to expect food two minutes after the bell was rung. Eventually the dog's mouth began to water two minutes after the bell was sounded— whether food was offered then or not!

One of the most interesting experiments was carried out by Ruch with white rats. The apparatus consisted of a box with a door leading to an alley. Both the box and the alley could be electrified separately, so that if the rat stayed too long in the box or ran too quickly into the alley it got a slight shock. The object of the experiment was to see to what extent the rat could learn the safe and unsafe time intervals.

The time between putting the rat into the box and the *middle* part of the interval of safety (i.e. when neither box nor alley was "live") was kept constant at 438 seconds. During the experiments the length of the safety interval itself was started at 292 seconds and then progressively reduced. It was found that the shortest interval of time the rat could learn was 56 seconds, or 13 per cent. of the standard time of 438 seconds.

This description of Ruch's experiment is, I am afraid, necessarily rather complicated, but it becomes clear if an actual trial run is envisaged.

Suppose the rat was placed in the starting box at twelve noon. The alley would be electrified and the box safe until six minutes and five seconds past twelve. From then, until seven minutes one second past twelve, both box and alley would be safe. After that the box itself was electrified. The rat learned to wait in the box until six minutes five seconds past twelve and then, before seven minutes one second past twelve, to run along the alley.

The time-sense of several species of insects has also been investigated under laboratory conditions. Grabensberger (1933) trained ants to visit a feeding-place at a given time each day. Within three days of the beginning of the training period, the ants learned the appointed time for feeding.

It was discovered that the ants could be trained not only to a feeding periodicity of twenty-four hours, i.e. to come for food at the same time each day, but also at intervals of three, twenty-two and twenty-seven hours. Once such a feeding rhythm had been established, it persisted for six to nine days after the experimental feeding had been discontinued. Grabensberger was successful in training termites in similar tests.

He also found that by upsetting the metabolic rate (i.e. changes in the living cells of the body) of the ants, their time-sense could be affected. When their metabolic rate was increased, by feeding them on thyroglobulin or by raising the temperature of their nest, the ants came for food before the training time. When, however, the metabolic rate was reduced, by feeding the ants on euchinine or by lowering the temperature of the nest, the tendency was for the ants to come for food later than the training time.

As far as I can gather, the greatest number of experiments on the time-sense have been carried out on bees. Von Frisch put out dishes of sugar and other foods attractive to bees at the same hour each day. Bees that visited this open-air cafeteria were caught and marked with tiny spots of red paint. Each day new visitors were caught and removed, so that only a definite group of bees could acquire the habit of coming regularly to the meals. When a time-keeper was stationed to record the arrival of the bees, he found that a large majority arrived at almost precisely the time at which the food had regularly been set out.

When no food was put out the bees came at their usual mealtime. When food was not put out for a week, the bees still turned up promptly at the time they had learned to associate with free food.

In Wahl's experiments (quoted by Washburn) bees were trained to come for food at six distinct periods, but the time interval between each meal had to be not less than two hours. The bees learnt to distinguish between the time intervals for two feeding stations only seventeen yards apart. Incidentally, Ingeborg Beling (1929) found that bees quickly learn the feeding times to which they are trained. In some experiments, bees came to the feeding-place at the specified time the day after training began.

She also found out two interesting facts about the nature of the bees' sense of time. They appreciate points but apparently not periods of time. The unit of their time-sense is the twenty-four-hour period. She bases these assertions on the following facts.

She trained bees to come for food every twenty-four hours and every twelve hours. This meant that they came at the same time, or times, each day—say 7 a.m., or 7 a.m. and 7 p.m., each day. She then tried to teach the bees to come for food every nineteen hours. Now this meant, of course, that while the period remained constant the actual point of time varied each day. And in this experiment the bees failed completely, as they also did when a forty-eight-hour time unit was used. Hence the conclusions that bees can appreciate points but not periods of time, and that their time-sense is closely connected with the twenty-four-hour rhythm.

The observations of Kleber, who has dealt with the biological significance of the bees' time-sense, throw an interesting light on these facts. A study of thirty-five varieties of flowers showed that there were daily rhythms in pollen production and in nectar secretion. In eight plants pollination was found to begin and end each day at a constant time.

Kleber kept a record of the visits paid by bees to flowers of different varieties. He found that these visits were made at times which corresponded closely to the secretory period of the flowers and that visits were most numerous when secretion was at its maximum. The bees visited flowers in great numbers at the time of the maximal quality (i.e. sugar-content) rather than at the time of the maximum quantity of secretion.

The significance of Ingeborg Beling's findings will now be apparent. A twenty-four hour period has a biological significance for the bee, as some flowers have a daily rhythm of secretion. And under laboratory conditions bees can be trained to respond to similar points of time within the twenty-four-hour period. But periods of time, divorced from the twenty-four-hour period, have no biological significance for the bee, and it has, therefore, proved impossible to train them to respond to such units of time.

The above facts having been established, the next thing to be discovered was what it is that serves the bee for a clock. It cannot be hunger, because bees eat food in their hives at all hours. Neither can it be outside influences, such as the position of the sun, or state of temperature, humidity, or illumination, for experiments have been carried out in laboratories where daylight was excluded and other factors were constant, and still the bees kept to the time schedule.

It was thought at one time that daily changes in the intensity of the cosmic rays reaching the earth might act as the bees' clock. So all the experiments were repeated at the bottom of a salt mine, where cosmic rays cannot penetrate and where, as far as present knowledge goes, all external factors which might give a clue to the time of day are excluded. But still the bees learnt to come for food at the proper time! (Fox, 1940.)

Calhoun points out, however, that all the above experiments were carried out with adult bees "which had had the opportunity to establish an endogenous rhythm, which no doubt would affect the results of any feeding-training programme."

At present the most widely accepted theory of the origin of the time-sense in bees, and in other animals as well, is that it is connected with the metabolic rate. As in the instances of ants mentioned earlier, it has been found that artificial disturbance of the bees' metabolic rate upsets their time-sense. Kalmus (1934) trained bees to visit a feeding-place for food at the same time each day. Then the bees were subjected to various conditions affecting their metabolic rate. When they were kept for six hours at a temperature of 50° C., the average delay in coming to the feeding-place was two and a half hours. Carbon dioxide narcosis and feeding with euchinine also affected the bees' time-sense.

Kalmus also subjected the bees to etherization for several hours, prolonged darkening of the hive, and he also fed them food containing thyroid extract. None of these conditions, however, affected the general accuracy of the bees' sense of time.

Grabensberger (1934, *a.*) carried out similar tests, using both bees and wasps. He found that feeding with euchinine made the insects feed later than the training time, but thyroiodine made them feed earlier. The influence of the drugs was less marked with the wasps than with the bees.

"This shows that the temporal memory has an endogenous, i.e. developed internally basis. However, since application of chloroform and of ether over a period of hours had no apparent effect upon any learned rhythm, the factors underlying this phenomenon must be non-nervous in nature."

In later experiments, Grabensberger (1934, *b.*) found that small doses of either salicylic acid or yellow phosphorus made ants come for food earlier than the training-time—the yellow phosphorus causing the average of visits to be six and a half hours earlier!

8

The Mystery of Bird Anting

ONE OF the ornithological mysteries is why birds use ants as part of their toilet. Whatever the explanation may be, the habit has been attested too often by ornithologists in four continents for there to be any doubt now of the occurrence of the phenomenon. I say "now" advisedly, because it is only within comparatively recent years that the habit has been generally recognised as a piece of authentic ornithological lore.

How comes it then, that although the phenomenon had undoubtedly been witnessed by observers of birds, including experienced ornithologists, for a hundred years or more, until the last ten years or so one could have searched the bird literature of the world and found hardly a mention of the habit? Perhaps the observers of the habit did not fully appreciate the significance of what they saw, or they did not know that they were privileged to see a rare and unusual rite which had hardly ever been recorded. Or was it, in some cases at least, that the observers could scarcely credit their senses and were afraid of being ridiculed for publishing what they thought they had seen the birds doing? (See, for example, Van Tyne's remarks on pp. 165–66.)

As far as I can gather, and I think I have either read or seen references to most of the statements which have so far been published on the habit, two well-known ornithologists are mainly responsible for putting this interesting

habit on the ornithological map. The first man to take a
serious interest in it was A. H. Chisholm, the well-known
authority on Australian birds. The following is his own
account of how his attention was first roused.

> A small boy living in a Melbourne suburb wrote me that he had
> seen starlings actually picking up ants and stowing them beneath
> their wings. Frankly, I doubted the evidence of the boy's eyesight.
> But it chanced, just afterwards, that I was looking through some
> notes I had written years earlier, and among them I came on a
> paragraph, long forgotten, to the effect that a man in Sydney had
> once seen certain soft-billed birds placing live ants beneath their
> wings.
>
> That started me on a search through the bird books and journals
> of both Australia and Britain, and of America as well. Finding
> nothing on the subject in any of them, I began to wonder if the
> practice was confined to Australia, perhaps because of the nature
> of certain ants there. Anyway, I mentioned the matter in a book
> [*Bird Wonders of Australia*—1934] soon afterwards. Then events
> began to happen—and they have been happening ever since, in
> four continents. There have been, so to speak, international com-
> plications; and all because of that boy who spied on the star-
> lings! (Chisholm, 1940.)

That Australian boy, Peter Bradley, who had the good
sense to write about what he had seen to a well-known
authority on the birds of his country, surely deserves his
niche of fame among the original discoverers of the life
histories of the birds.

Shortly after Chisholm's book was published, Professor
Erwin Stresemann, of the University of Berlin, quoted in
a famous German bird journal what Chisholm had written,
and asked anybody who had observed similar occurrences
to let him know. The result of Stresemann's inquiry was
remarkable. He received letters from all over Germany
supplying most valuable information on the habit, and re-

calling half-forgotten notes which had already been published in German journals.

Stresemann published the result of his inquiry and suggested the use of a special term, "einemsen." This has been translated into English as "anting," and it is now generally used to describe the habit, even when other insects, or even non-animal substances, are used in the process.

Several typical examples of anting which have been put on record by reliable observers will indicate the nature of the habit.

Charles K. Nichols, of the American Museum of Natural History, says he saw on his lawn a robin going through some remarkable actions. The bird picked up something from the ground and then quickly placed it under one of its partly opened wings and sometimes on the under-side of its tail. Frequently the bird lost its balance and fell on its back. In addition to these actions the bird sometimes pressed its breast to the grass and partly rotated its body with the breast as a pivot.

Later another robin appeared on the scene, drove the first one away and after settling on the same spot on the lawn went through the same sort of actions as the first bird. The second robin was in turn replaced by a third, which occupied the favoured spot for a few minutes. Thereafter the birds took "turn and turn about." When Nichols examined the spot, while the birds were momentarily frightened away, he found a swarm of about a hundred ants milling excitedly about a space a foot or so square.

Josselyn Van Tyne, the editor of *The Wilson Bulletin,* prefaces the following record of what he saw by the statement: "I never fully believed in the occurrence of this most improbable phenomenon [bird anting] until I recently saw it with my own eyes."

Van Tyne says that soon after sunset one evening in July he saw a male robin preening itself on the lawn fifteen feet from his window. He continues:

> The bird was preening much more vigorously than is customary and his actions were further remarkable for the frequency with which he preened in a single motion the whole outer edge of the wing from wrist to tip. In fact, this wing preening was done so violently that the bird repeatedly fell down at the end of the preening motion and once this ended in a complete somersault.
> Sometimes the bird preened the tail or body plumage, but more often he concerned himself with the wing. Almost immediately I noticed that nearly every preening was preceded by a hasty picking of some small object from the ground and I realized that here at last was a bird "anting." Several times the robin crouched and seemed to rub its body against the ground.

As the following observations show, there are several ways in which birds ant. The first concerns two tame jays which used to fly about a farm. Whenever an ants' nest was laid bare in the course of the farm work the two birds trod on it, and this stimulated the ants to shower the birds' feathers with their acid ejections. Sometimes the jays wallowed in the nest. They often raised their tails and sat down and then almost immediately turned on their sides. The jays stayed on the nest for periods up to a quarter of an hour. They then flew away and shook and preened themselves as after a water bath. (Robien.)

Some birds appear actually to bathe in ants. A pair of starlings were observed to bury themselves in an ants' nest and then to throw the ants over their feathers with every sign of enjoyment. The birds were seen also to pick up ants with their beaks and place them under their feathers. (Floericke.)

Similarly a tame carrion crow has been seen to bathe in swarms of ants, one such bath lasting for twenty-five minutes. This bird was also seen to gather a number of ants in its bill, squash them, rub them through its plumage, cast them away in the form of a pellet, and then gather more. Very similar observations have been made with thrushes and other birds. (Ali.)

After I had published a short article on anting in *Country Life*, a R.A.F. corporal, P. G. Lewington, wrote from India saying he had witnessed a similar occurrence. As his account is the first I have seen of birds anting in trees, I quote the salient part of his letter here. He writes:

> In the jungle of Assam a species of large red ant make their nests in the trees. A few weeks ago a friend of mine and I were walking through the jungle when we were showered with ants. On looking up we saw three birds preening themselves with ants, one of the birds being a lovely parrot. We were not more than six yards from them. They were in the same state of ecstasy described by Mr. Lane, so much so that they took no notice of us at all. We watched them for several minutes. They picked up several ants at once, and rubbed their wings and feathers. This happened early in the morning.

To these and a number of other existing records of observations of anting, there have been added two accounts of the habit witnessed under controlled conditions, by Adlersparre in Sweden, and Ivor in Canada. Ivor (1943) scattered a shovelful of earth, containing several hundred ants, over part of the floor of his aviary. He then lay down on the ground close to the birds (within sixteen inches at times!) to watch their reactions. Some of them actually anted on his hand.

Sixteen experiments were carried out altogether, and twenty out of the thirty-one species of birds in the aviary were observed to ant. About a score of other species have been reported by other observers to ant. Writing in 1946, Ivor says: "Perhaps half of our passerine birds indulge in this extraordinary performance at times," and, referring to his own observations of the habit, he says: "There seemed to be no fundamental differences in the specific actions of families, species, or individuals, the only variations being in position."

Ivor (1943 and 1946) thus describes the performance as he witnessed it in his aviary at close quarters.

> The moment an ant was sighted by any bird which anted, there seemed to be an instantaneous and instinctive reaction. The ant was picked up and held in the tip of the bill; the eyes were partly closed; the wing was held out from the body, but only partly spread; the wrist was drawn forward and raised, thus bringing the tips of the primaries far forward and touching the ground; the tail was always brought forward and under to some extent, on the same side as the extended wing, and often so far that the feet were placed upon it. Stepping on the tail at times caused the bird to fall on its side or even on its back. . . . The performance is so extraordinary, so foreign to any other behaviour of birds, and so clownishly beautiful that it almost brings tears of laughter.

The bird which seizes the ant rubs it swiftly only on the ventral, or under surface of the primary wing feathers. The performance is so rapid that it is difficult for the eyes to follow it, and Ivor thinks this is the probable cause for the great divergence of opinion among observers as to what exactly does take place.

Ivor never saw ants placed among or under the feathers, or rubbed on the legs. After being used for anting, the ant was often, though not always, eaten. Other observers confirm that ants are sometimes eaten after being used for anting, but this is by no means an invariable practice.

Ivor (1943) noticed that enthusiasm for anting varied with the season. He adds:

During the height of the anting season the act of anting seemed to engender a state of ecstasy so overwhelming that even domination and enmity were forgotten. The rose-breasted grosbeaks (*Hedymeles ludovicianus*) are very quarrelsome, but it was rare to see even one of these make a belligerent movement toward another bird during the performance. This, too, in spite of the fact that, at times, from twenty to thirty birds would be going through the performance at one time on a space of four to five square feet, where they were continually bumping against one another.

Confirmation of Ivor's belief that anting is instinctive is found in the reactions of young birds when first confronted with ants. Young starlings, taken from the nest and given some ants, dressed their plumage just as in the case of adult birds. (Chisholm, 1940.) A young dipper, when first presented with some ants, seized one after another in its beak and passed them through its feathers. (Heinroth.)

Ants are not the only things to be used in anting; other insects and non-animal substances are used at times. Hampe says tame starlings, which inserted ants among their feathers at every opportunity, would also eagerly dress their feathers with the juice or flesh of a lemon, vinegar or even beer. He says they were always keen to find and bathe in a bowl in which salad had been prepared with vinegar. A tame jay had a similar fondness for orange juice. Every time an orange was peeled this bird came near to intercept the spurting juice and went through the motions of bathing.

A correspondent of Chisholm's (1944) recorded an unusual trait in his tame magpie. It would fill its beak with ants from the garden and then enter the house and find someone who was smoking a pipe. The magpie then flew on to the pipe-smoker's shoulder, dipped its ant-filled beak into the hot ash and then put the resulting mixture of ants and tobacco ash under its wings! The writer said: "I used to try smoking him out, but while there was any ash left in a pipe he would not leave, except to collect more ants." Heinroth writes of a magpie which eagerly rubbed its feather with cigar stumps.

Other substitutes for ants which have been noted by various observers are beetles, amphipods, mealworms, bugs (*Rhynchota*), the rind of limes, sumach berries, choke-cherries and even pieces of apple and apple-peel. Groff and Brackbill repeatedly watched flocks of purple grackles anointing themselves with juice from the husks of English walnuts, and Hill watched a bronzed grackle anting with mothballs.

Nice and Pelkwyk quote from an entry in their bird diary as follows concerning a song sparrow "Y." "Y is on the floor, 'anting' himself with a sumach berry; he hunches about in a very queer attitude. He has lost the inner rectrices. He is evidently trying to rub the berry on his feathers. He turns about, nearling sitting on his rump—a most ridiculous position—and tries to rub the berry on his tailfeathers. He then goes up on top of the sumach branches and tries to rub berries under his wings."

Another correspondent of Chisholm's (1944) said her pet parrot had the habit of taking a piece of apple, or a long string of apple-peel, and pushing it beneath its wings or into its back between the shoulders. These actions were not performed with any other type of food. A cockatoo has also been observed rubbing itself with apple-peel. Incidentally, it has been proved that apple-peel has a cleansing effect on human beings. Dr. Charles McLaren, while a prisoner of the Japanese, found that in the absence of water, he could keep himself clean by rubbing himself with apple peelings. They also made "an excellent and very agreeable dentifrice." (*Victorian Naturalist,* September, 1943.)

Givens records a remarkable example of anting he observed in North Queensland, when a small flock of red-browed finches daily visited a smouldering log and used the tendrils of acrid smoke curling up from the cracks in the bark as an anting medium. He writes:

> The same routine was followed each day. First one and then another of the birds would fly on to the log, until the whole flock of about a dozen or so birds was present. . . . Once on the log the actions observed were quite distinct from those commonly seen when birds bathe in dust or water. Upon reaching the smoke, each bird stood as upright as possible, using its tail as a support. The wings were opened slightly and drooped a little forward and down. Then the head was swept forward, down and under the wing in a circular motion, the bird meanwhile vigorously shuffling its wings and body feathers, often toppling backward from the violence of its efforts. These actions were rhythmically repeated as many as eight or ten times, when the bird usually rested before repeating the whole process.

Givens was able to watch the performance from a distance of not more than six feet and he is certain that none of the birds picked up anything from the log. He examined this afterwards and found no ants or other insects on or under the bark. On one occasion, when a piece of bark was torn away from a nearby stump, and an ants' nest was disclosed, one of the birds tried to ant there, but it soon abandoned the attempt in favor of the smoke. Although in the thick of the smoke, none of the birds appeared to be as disturbed by it as Givens himself, some feet away, although he noticed the nictitating membrane flicking across their eyes more than usual.

He noticed one other interesting example of bird behaviour during this performance. On several occasions one or two of the birds could not find room in the smoke, but they went through the motions of anting a few feet away! This behavior is similar to that of birds which, unable to find room in a bird bath, will go through the actions of bathing on the ground.

Why do birds ant themselves? Various theories have been propounded. Perhaps the most remarkable was that the ants were stowed away among the birds' feathers to act as a sort of portable pantry during their migration flights! This theory is generally discarded today.

Another suggestion is that the bird is stimulated by the crawling of the insects, their tiny bites and acrid secretions. This pleasure may be akin to that derived from the ruffling of a bird's feathers by a human hand, an action enjoyed by numerous birds in captivity. Some writers have thought that the ants may have a stimulating effect on the birds' skin.

Another theory is that birds resort to anting to rid themselves of parasites. The formic acid secreted by ants has antiseptic properties. Birds have been seen to hold ants in a way that would indicate the bird was trying to make the ants spray their acid on the feathers.

Frazar says he has seen ants seize the parasites on a crow which was anting and bear them away. In this connection it is interesting to learn that in some parts of the world ants are sometimes used to remove vermin from clothes. The infected garments are placed on large ant hills and, when collected, they are found to be freed from vermin (see Wheeler, and Chisholm, 1944). Hodge says when he was in Canada he saw a Frenchman take off his shirt, lay it on the ground and let ants run over it. "When he judged they'd had sufficient time to eat the lice, he shook out the ants and put it on again."

Chisholm (1944), who reviews the whole subject and the various theories which have been propounded, comes to the following conclusions. He says:

I regard the evidence as indicating that "anting" is practised by birds, according to circumstances, because of (*a*) the value of the acid as a skin stimulant; (*b*) the consequential value of acids when applied to feathers in cleansing the body of parasites; (*c*) the appeal of the odours of formic acid and other pungent mediums; and (*d*) the desire to free the ant of acid before eating it, or, more probably, the desire to enjoy external benefit as well as the benefit afforded by the ant as food.

One other apparent use of anting was mentioned long ago by Audubon, in fact, it appears to be the first reference to the habit. He says young Eastern turkeys "roll themselves in deserted ants' nests to clear their growing feathers of the loose scales and prevent ticks and other vermin from attacking them, these insects being unable to bear the odour of the earth in which ants have been."

Ivor (1946) considers that none of the above explanations are convincing. He sums up his own views as follows:

> So far, after intensive investigation, I am unable to make a single plausible suggestion as to the biological significance of this exceedingly peculiar behaviour, and I am compelled to agree with McAtee, who states: "The phenomena involved in anting . . . are both remarkable and obscure, and whether we shall ever understand their exact significance is doubtful."

H. Roy Ivor, who has probably watched more examples of anting than anyone else and is the foremost authority on the subject, has kindly sent me the following comments on this chapter:

"I have seen thousands of reactions among song-birds, and, although I have not been able to watch all of them, not one of those I observed closely showed any evidence whatsoever in support of the suggestion that birds resort to anting to rid themselves of parasites. In each performance, the ant was rubbed on the ventral surface of the primaries (perhaps some on the secondaries, although this is doubtful), beginning about two-thirds from the wrist and continuing to the tips, or, rarely, on the ventral surface of the tail. My article in *The Auk* stated that the ant was rubbed on the primaries from the wrist to the tips. That this statement was an error was shown by slow motion pictures, which demonstrates how easily an error can be made in reporting what is actually seen.

"It is not surprising that similar mistakes have probably been made by other observers: in the first place, the performance is extremely rapid; secondly, there is the surprise element in viewing a behaviour hitherto unseen; lastly, one has to reckon with the human failing of not seeing accurately what really happens. It is a well-known fact that where half a dozen people have seen an accident, no two will describe it in the same way.

"Until now, in the majority of cases, inferences as to the significance of anting have been based upon viewing, from a distance, a performance by a wild bird, or birds. Obviously this does not lead to accuracy in reporting, and I think that such evidence should be largely discounted in determining the significance of anting. To my mind, the only reliable test is one which can be made with birds sufficiently tame to allow of close and continued observation. This does not mean that my experiments have proved finally that there is only one way in which a song-bird ants. What I have proved, however, is that the many birds which I have had under observation for several years have a definite pattern, which is never altered.

"This does not prove, of course, that all song-birds react in the same way, but it forms a basis for inferring that it is a 'parent' pattern, which probably relates to most, if not all, song-birds. I think that if we are justified in deducing theories in regard to significance, such inferences should be drawn from experimental work. With the exception of the crow, which I shall mention later, my experiments have given a definite pattern of behaviour so baffling that I am unable to deduce any general principle.

"We may expect one bird to act differently from the rest of its species, and therefore when Nichols reports how an American robin pivoted its body among ants, I think that we must take this behaviour to be an idiosyncrasy. My own robins have performed hundreds of times and not once have I seen them vary the set pattern. Of course, the spraying ants may stimulate a bird in other ways than would those which bite.

"Besides the numerous experiments I have made, I have had taken about two hundred feet of normal-speed, coloured action pictures of birds while anting, in addition to fifty feet of coloured, and fifty feet of black-and-white, slow motion pictures. I have scrutinised many of these frames through a viewer and have had several of them printed and enlarged. Not once have I been able to discern any variation in the pattern. I have used over one hundred birds and about forty species in the experiments. These motion pictures do not show the bird placing the ant on any part of the plumage other than the underside of the primaries and the tail.

"My starlings (*Sturnus vulgaris vulgaris*) performed exactly similarly to the other species which anted. The only exotics which indulged in this behaviour were the European blackbird and the Asiatic Pekin-Robin (*Leiothrix lutea*), and their performance followed the same pattern as that of the native birds.

"When we consider the problem of anting, we are faced, according to my experiments, with the fact that two parts of the plumage only are anointed. With the exception of the head, these two parts are obviously the most inaccessible to the bill of the bird. This accounts for the contortions of the body and, in my opinion, precludes the possibility of accepting any of the suggestions previously made as to the biological significance of anting, with the possible exception of pleasure derived. Even this, however, does not seem logical, for, were it done for pleasure, there are many parts of the body more readily accessible. This applies to Chisholm's four points, as well as to those of others.

"The fact must also be faced that while some species eat ants, they do not ant. It can be safely assumed that the species which do not ant are just as subject to the various supposed desires as are those which ant. The bluebird and the American robin, which are, anatomically, supposed to be closely related, illustrate this point. The former nests in a cavity and should therefore be more subject to ecto-parasitic infestation than the robin: yet the latter ants, while the bluebird does not.

"Let us consider the bluebird. Does it not need an acid to cleanse the feathers of parasites? Does it not need a skin stimulant? Do the odours of formic acid not appeal to it? Does it not desire to free the ant from formic acid before eating it? It seems remarkable that the bluebird does not need the ant for any one of these purposes and the robin does.

"I have not included the crow or the magpie among the song-birds, although they are passerines. If the crow is included, we have a departure from the pattern common to all the other birds with which I have experimented. Occasionally, the crow places ants among her feathers. She sits among the ants and, holding her wing tips crossed in a normal way, extends the wrists from the body to some extent, holds them in this position, and allows the ants to crawl over her. I have never seen this pet crow 'ant' or perform in the same way as other birds. She does not place the ants on the underside of her wings and tail and, whether or not the ants penetrate the feathers on the underside of her body, I am unable to discern. Certainly the ants which I could see did not do so. Some ants, however, seized the tip of a feather at times and tried to bite or pull it. Evidently the crow derived great pleasure therefrom, for she closed her eyes and remained quiet for minutes at a time. I believe that, as she enjoyed my stroking her head and back with my hand, so did she like the feel of the ants crawling over her. Other birds will not allow me to stroke their feathers and, judging by the way they try to pick off ants which crawl over them, they do not enjoy this sensation either."

Regarding the suggestion that anting in aviaries may not be typical of anting in the wild, Ivor says:

"I do not believe this is necessarily correct. In my view there is, in all probability, more anting among wild birds than is known. As far as I have seen, a liberated aviary bird may, when it has the opportunity, ant only two or three times. It is because birds in an aviary cannot, as wild birds can, ant when they please, that when they do have the chance, their performance is more noticeable. For a time they will ant with great enthusiasm, then they grow tired of it: consequently, were I always to keep ants in the aviary, I doubt if I would see more than occasional performances. Further, when, during the spring, I trapped an olive-backed and a grey-cheeked thrush, it was approximately two weeks later before they anted. Certainly they could not, in that time, get in such conditions as Chisholm suggests."

9

The Legend of the Hedgehog and the Fruit

THERE ONCE came into my hands a remarkable photograph taken by an anonymous European photographer. (See plate 54.) It was a flashlight photograph of a hedgehog apparently carrying on its spines two pears and some leaves. As soon as I saw the photograph I remembered the ancient legend concerning the hedgehog carrying fruit on its spines.

Nearly two thousand years ago Pliny wrote in his *Natural History* (*circa* 75 A.D.): "Hedgehogs lay up food for the winter. Rolling themselves on apples as they lie on the ground, they pierce one with their quills and then take up another in the mouth, and so carry them into the hollows of trees."

From Pliny's day to our own this story has cropped up from time to time. The Hedgehog and Fruit Legend, as it may be called, was a favourite subject with the compilers of the bestiaries and other finely-illuminated manuscripts of the mediaeval ages. (Druce.) The legend figures also in the armorial bearings of several English families. Moreover, in Lincolnshire, England, there is (or was in the last century) a saying for a man with a ruffled temper: "He has gotten his back up, like a hedgehog going 'crabbing.'"

The legend is also referred to in John Clare's poem, *The Hedgehog*, in which he says:

> "The hedgehog hides beneath the rotten hedge
> And makes a great round nest of grass and sedge.

Or in a bush or in a hollow tree;
And many often stoop and say they see
Him roll and fill his prickles full of crabs
And creep away; . . .

The Hedgehog and Fruit Legend may, in fact, be regarded as one of the standard controversies of natural history. It must be frankly admitted that the majority of recent writers on the hedgehog who have mentioned this story have taken the negative view. Representative of such opinion is Frances Pitt. In the following passage she firmly denies the truth of the story:

> With regard to this legend, we must remember the animal's insectivorous and carnivorous tastes—it has no vegetarian inclination whatever. Nevertheless, the yarn is that when, in autumn, the apples come tumbling from the trees, hedgehogs visit the orchard, roll upon the fruit, and with apples affixed to their prickles return to their homes. The only foundation for this impossible story is that sometimes, when an urchin [the hedgehog] comes forth from its resting-place, it may have a bit of grass, or a piece of leaf, caught on its armour.

On the other hand, Miller Christy, who thoroughly investigated the literature on this subject and to whose researches I am indebted for much of the information in this chapter, wrote, after summarizing and criticizing the evidence he had collected:

> The conclusion I arrived at was that, although [the] evidence did not suffice to prove, directly and incontestibly, the truth of the legend, it was, nevertheless, sufficient to leave no reasonable doubt that the legend, however improbable on the face of it, was really justified by observed facts.

In passing, I should like to say that I consider no one, however eminent he may be as a field naturalist, has a right to dismiss this subject as unsubstantiated legend unless he has read the evidence in favour of the story which Christy collected and collated with considerable industry, and published with full documentation.

It is of interest that both Christy's articles were communicated by T. A. Coward. I find it difficult to believe that such a trustworthy naturalist would have lent the authority of his name to a story which was as patently absurd as some opponents of the legend would have us believe.

Before citing the evidence in favour of the truth of the legend it is necessary to deal with the objection that the hedgehog is almost exclusively insectivorous and, to a small degree, carnivorous, and that it is never a vegetarian. The interested reader is referred to the early volumes (1843, 1844, etc.) of *The Zoologist,* where this question was fully debated. Some writers said they had kept tame hedgehogs which had definitely eaten vegetable substances, including apples. The majority of writers, however, maintained that captive hedgehogs do not eat vegetable fare, but, as Christy points out, the tastes and habits of captive animals are not necessarily the same as those of wild ones.

In a personal communication to Christy, James E. Harting wrote:

> From personal observation I know that they will feed on fallen fruit. On one occasion, late in September, I was returning home in the evening and saw at a little distance a hedgehog at the foot of a crab-tree, busily engaged in mouthing some object which, at the distance, I could not distinguish. I watched it for some time and then, as I slowly approached, the hedgehog scuttled away. At the spot I found a partially-gnawed crab-apple bearing the marks of teeth on one side, which convinced me that the animal had been feeding on it. So far as I could see, the hedgehog made no attempt to carry off the crab in its mouth, as a squirrel would have done.

It is only fair to add that Harting did not believe the Hedgehog and Fruit Legend.

"The King of the Norfolk Poachers," writing out of a wealth of experience of English animal life (though not of English spelling!), says in that enchanting book, *I Walked By Night*:

> If put to it he [the hedgehog] is a vegetairan as well, and will even eat crab apples when he can find them. I have dug out half a peck of crabbs from rabbit burries and other holes were he have laid them up for the winter.

I consider the above evidence, taken in conjunction with other statements to the same effect quoted by Christy, establishes the fact that, on occasion, the hedgehog will eat vegetable fare. Before leaving the subject, however, I must mention an interesting suggestion put forward by Spicer. He believes that the hedgehog is a vegetable feeder only, or mainly, when young.

Writings on Chinese hedgehogs, Ch'eng-chao Liu has the following passage: "Hedgehogs are suspected by farmers of eating cabbage, sweet-melons, water-melons and jujube 'dates.' A hedgehog is said by farmers to be able to carry a large number of jujubes to his den by rolling his body over the fruits that have fallen to the ground, so that a number of them become attached to the spines."

Is there any evidence that the hedgehog ever uses its spines to impale objects deliberately? I think there is. Christy says: "Anyone who has seen a dormant hedgehog taken from its hybernaculum must have noticed (as I have done many times) that the dry leaves with which it lines its nest are pierced by its prickles to an extent which cannot be wholly the result of accident."

Briggs also writes: "When found in winter they [hedge-

hogs] are encased in a coat of dry leaves, about half an inch
thick, which adhere to the prickles in so firm a manner that
the animals seem to have rolled themselves amongst them."
And in a fight between a hedgehog and a rat the hedgehog
was seen to endeavour to pierce the rat with its spines.
(Parker.)

More to the point, however, is an observation of M. Forster Knight's, who says she has seen a pet hedgehog transfix
food on its spines. The observation is so important that I
give it in full in her own words:

> I took the hay out of the box to wake him up, and caught some
> beetles and placed them under his chin. He ate them with relish:
> nothing, however, would induce him to round them up on the
> floor. When he had eaten three he twisted his head and impaled
> the remaining insects on to his bristles. I have never known an-
> other hedgehog do this. Whether he meant to store a meal or
> merely liked the idea of beetle decorations I do not know.

Before giving the direct evidence on which belief in the
legend is based it must be pointed out that in the nature of
the case observation is bound to be difficult. The hedgehog
is largely a crepuscular and nocturnal animal. It therefore
seldom stirs abroad until light conditions are very poor for
observation, so that even if the habit mentioned in the legend were common, witnesses to it are bound to be few, and
of those the proportion of trained and reliable observers
would be very small.

The first observation is that mentioned by Christy himself—in fact, it was this incident which originally caused
him to undertake his painstaking investigation into the
story. He was gathering crab apples from some of his trees
and had as his assistant George Franklin, an old farm-
laborer who had worked for him for several years. Christy

says he knew him well enough to accept his word in a direct statement with perfect confidence. Christy says:

> We had nearly done, when, finding we had gathered less than a bushel, I urged them to search further for other crabs which might have become hidden among the grass and bushes.
> "Why, master," says Franklin, "you should leave a few for the poor hedgehogs." "Hedgehogs!" said I, "what do they want them for?" "Why," replied he, "they eat them. I once saw one carrying some home on his back."

Christy questioned the old fellow and Franklin told him he was once walking home through the fields in the evening when, looking over a gate into a meadow, he saw, not more than three yards away, a hedgehog shuffling along with something on its back. Closer inspection showed that these were crab apples, evidently stuck on the little animal's spines. As the hedgehog was within a few yards of, and was coming directly from, a large crab apple tree, Franklin naturally concluded that the crabs had been obtained from beneath this tree.

The second observation was communicated to Christy by A. Hibbert-Ware. She was working in a museum when a Russo-Polish peasant came in and started to talk about the animals of the Russia he had left many years before. Just as he was leaving he caught sight of a stuffed hedgehog. He said: "How often, in Russia, have I seen those little animals walk away with apples or pears upon their backs."

Hibbert-Ware, who had heard of the legend, was naturally interested and asked her visitor for more details. He replied:

> They come to the apples lying on the ground, below the trees, and roll themselves up into a ball right on the top of them [imitating his meaning with his hands], and then they walk off with two or three sticking upon the prickles of their backs. I have

never seen one content to carry off a single apple only: they always have two at least.

A point to be noticed about both these reports is that there was no prompting which might have encouraged the narrators to fancify their observations. It is difficult to believe that either the old countryman who spoke to Christy, or the Russo-Polish peasant who told what he had seen to Hibbert-Ware, were recounting anything but what they had seen with their own eyes.

Mabel Peacock, writing from Lincolnshire, where the saying mentioned on page 178 was current, says: "I was told by an intelligent working-man that, when he lived at a situation in north-east Lincolnshire, he saw one of these animals [hedgehogs] carrying apples [on its spines] several times, in his employer's orchard. A pair of hedgehogs inhabited the place with their young; and, not only he himself, but other people, used to watch the old ones transporting apples on their prickles."

After I had published an article on this subject in *The Field*, Heathcote sent a letter to the editor in the course of which he said:

> About fifteen years ago my gardener, Mr. H. Howard, of Easton, was walking in the early morning across a meadow, when he saw a hedgehog under a crab-apple tree rolling on the fallen fruit. Mr. Howard waited until the animal moved away into the hedge carrying apples spiked on its back. When I showed him *The Field* photograph [plate 54] he said: "Yes, just like that, only more apples."

It may rightly be objected that none of these observations was made by a scientist or a trained observer, and that while the observer's *bona fides* are not questioned it is possible, as all scientists well know, for the untrained man easily to be mistaken in what he thinks he sees. The same objection can hardly be maintained in the case of the next observation. The incident is put on record by W. H. Warner, a member of the Selborne Society and a reliable field observer. He writes:

> I well remember many years ago meeting with hedgehogs in an Oxfordshire orchard, to the spines of two of which several apples were sticking. The apples had adhered to the spines, there was little doubt, when the creatures were rolling under the trees. That the hedgehog climbs the apple-tree and carries off the fruit stuck to its spines (as country people say it is in the habit of doing) is, of course, absurd.

Warner's observation was put on record in reply to a question from Doveton, who had written in the previous issue of *Nature Notes*: "Can any of your readers tell me whether it is the practice of hedgehogs to throw themselves inverted upon apples lying on the ground, so as to impale them with their spines, and then to rise and carry them off on their backs? . . . My gardener declares that he has *seen* the feat performed in an orchard hard by!"

The next account is taken from the German periodical *Die Gartenlaube*. It appeared in a translation in *Science Gossip*. The observer, "B.L.," of whom, unfortunately, details are lacking, said he had a pear tree in his garden and one morning he observed a hedgehog approaching it. "B.L." says:

> After snuffling among the fallen fruit for a short time, the hedgehog took one up in its mouth by the stalk, carried it a few yards, and laid it down carefully. It then returned, seized another and laid it by the first one. This was repeated until no less than sixteen pears were lying close together, or, rather, heaped upon each other. Satisfied, I suppose, with the amount, and conscious that it had collected as many as it could conveniently carry on one trip, the

animal spread out its prickles to their widest possible extent, deliberately threw itself upon the heap of pears, and rolled from one side to the other, until the whole of the fruit was transfixed. It then calmly walked off with its easily-gotten booty to the place from whence it had issued, where I could plainly see some little ones awaiting its return and, no doubt, anticipating a juicy breakfast.

Shortly after this account appeared in *Science Gossip* the editor, Mordecai C. Cooke, who had received additional accounts from people who claimed to have witnessed hedgehogs transporting fruit in this way, wrote: "In the face of four independent assertions of the fact, we think that the assumption is strongly in its favor, and that all who are still disposed to be sceptical must for a while suspend their judgment."

This editorial pronouncement was quickly followed by a communication from no less an authority than Charles Darwin. Here again I would point out that it is in the highest degree unlikely that a man of Darwin's stature would have taken part, on the positive side, in a controversy concerning a legend which is as incredible as some writers would have us believe.

Darwin's communication related to an observation of a Mr. Gisbert, who was in the Spanish Consular service. Gisbert said he had often seen hedgehogs in Sierra Morena moving along with at least a dozen large strawberries sticking on their spines. In this part of Spain such strawberries are very abundant.

Moll gives a detailed account of an incident he witnessed in Central Europe. His father reported seeing a hedgehog very early one morning rolling on pears beneath a pear tree on a steep hillside. The hedgehog then bore them away to a hole in a stone wall. Moll says:

Next morning my brother and myself accompanied my father to the orchard before daybreak. After some time our little friend appeared and repeated his performance. We saw him roll down at least six times, after which he had a fine load of about two dozen small pears. He then went off to his lair in the stone wall, but in less than five minutes he returned "empty." He appeared to know exactly where he had left off, as he next devoted himself to the succeeding rows, and loaded up once more. This time we followed him to his hole, but could not reach him.

Further accounts supporting the truth of the legend can be found in Christy's articles.

It may fairly be asked, assuming the truth of the legend, why should a hedgehog go to the trouble of transporting food in the way described? Why does it not eat the fruit on the spot? There are two possible answers to this question. First, there is some evidence (see, for example, the quotation by "The King of the Norfolk Poachers" on page 181, and the photograph of a hedgehog's alleged hide-out —plate 56) that hedgehogs store food before they hibernate.

The second answer is that the hedgehog transports the fruit to feed its young. In this connection, see Spicer's suggestion on page 181. Moreover, in the accounts from *Die Gartenlaube* and Mabel Peacock, young are definitely mentioned.

It may be remembered that, unlike most British mammals, the hedgehog produces litters in the autumn when, of course, apples are most abundant.

The question has been raised how, assuming the truth of the legend, does the hedgehog remove the fruit from its spines? If, as I have suggested above, the fruit is carried to the young the answer would appear to be that they pull it off.

In any event the weight of a hedgehog (they seldom ex-

ceed three pounds), part only of which would be pressed
against an individual fruit, would not be sufficient to drive
the spines in very far. A slight sideways knock, as against
the wall of the "larder," would probably suffice to dis-
lodge the fruit.

Spicer answered the question as follows:

> Whether she can herself remove the fruit from her back by a
> vigorous shake, or by passing her body under the stiff branchlets
> on the lower part of a hedge, I cannot tell. But I have no doubt the
> task of detaching it is easily accomplished by the teeth and claws
> of half a dozen hungry young hedgehogs; for, after all, the dorsal
> spines are very short, and would scarcely enter the pear beyond
> its rind.

Such then, briefly, is the evidence for accepting the
ancient Hedgehog and Fruit Legend. If the photograph
which aroused my own interest in the problem be accepted
as genuine, then I consider there is a case for accepting the
legend as a very interesting piece of authentic natural history.

As already mentioned, it comes from Continental Europe,
where the majority of the observations concerning the legend have been made. "One point which the foregoing observations seem to bring out clearly is that (assuming the
hedgehog really does transport fruit in the way stated),
it does so less commonly in England than in central, southern and eastern Europe." (Christy, 1919.) This is not the
first time I have noticed a difference in the habits of the
British and Continental races of the same animal.

I close with two quotations. Dr. E. W. Gudger, who
among other notable contributions to zoology, proved that
the once-scoffed-at story of fish falling from the skies was
authentic (see *Scientific Monthly*, December, 1929), writes:

In the course of years of study of animals and their behavior, and thanks to extensive bibliographic work, I have learned that animals do so many extraordinary and unexpected things, that I am not ready to cast overboard as without foundation some alleged behavior (not physically impossible).

Finally, the words of Christy himself:

But is the story really so incredible after all? Are we not apt, in these highly-scientific days, to become too contemptuously sceptical in regard to all ancient legends of the kind, and to forget that, however absurdly improbable they may appear at first sight, not a few of them have been shown to have some genuine basis in fact—often slight, but sufficient to substantiate and justify them.

In all such cases, a cautious scepticism should be, of course, maintained up to a certain point: but I have never forgotten a dictum to which I remember hearing the late Professor Huxley give utterance many years ago: "I have always felt (he said) a horror of limiting the possibilities of things."

In this connection see the hedgehog article in Brehm's *Dyrens Liv,* edited by Prof. Sven Ekman and published in Stockholm, Sweden.

10

Bird Hitch Hikers

FOR A NUMBER of years I have been interested in this question, and have made notes of references bearing on it which I have met in various works and in correspondence and conversations. One of the reasons which has made me hesitate to publish my findings is that many experienced ornithologists regard the subject as one with hibernating swallows, goat-sucking nightjars, and goose-hatching barnacles.

I should still be diffident about publishing anything on this subject were it not for the fact that no less an authority than Dr. W. L. McAtee, of the Fish and Wildlife Service, published, in March, 1944, an article on this subject in *The Scientific Monthly,* which is an official journal of the American Association for the Advancement of Science. He reaches a conclusion which, while by no means accepting the alleged habit in its entirety, definitely accepts the fact that birds sometimes ride on the backs of other birds.

Before McAtee, James E. Harting had written on this theme in his *Recreations of a Naturalist* (1906), and had by no means dismissed it as an example of nature-faking. In face of what these two authorities have written I am emboldened to present my own contribution to the subject. In order to give the evidence fairly fully I have included in this chapter much of both Harting's and McAtee's material.

In the first place it must be pointed out that there are really several forms of pick-a-back travelling by birds, and that it is only one of these which is seriously in dispute at the present day. The first concerns those birds which associate with various mammals and occasionally ride them pick-a-back. The Egyptian plover (really a courser) frequently makes use of the broad backs of crocodiles while it devours their parasites. Ox-peckers frequently consort with big-game, especially rhinoceroses, settling on the backs of these irascible beasts while pecking at the hosts of bots and ticks which find refuge in the folds of their wrinkled hides. There are several other examples of mammal-riding birds, such as buff-backed herons and cattle and big-game, egrets and elephants, sun bitterns and tapirs, cowbirds and bison, and several species of birds and domestic mammals. (McAtee, and Verrill.) While writing this chapter I saw a magpie perch on a cow. Incidentally, Oliver reports seeing numbers of birds (masked boobies?) at sea utilizing marine turtles as animate perches.

Seton records a remarkable instance of a cowbird surviving the winter by warming itself on a bison's back and sleeping in a hollow it had made in the beast's hair. These birds have also been known to alight on horses, mules, and even men.

What is more important from the point of view of this chapter is that there are several records by reliable men of birds of one species perching on the backs of birds of another species and sometimes travelling with them. McAtee says Herbert Friedman told him that ox-peckers sometimes alight on bustards and ostriches. The carmine bee-eater has been observed perching on Abdim's storks, ostriches and bustards, and has been seen riding the last two.

Myers writes: "I was delighted to see with my own eyes, on two occasions, a carmine bee-eater riding a great grey bustard. Once, when the bustard broke into a run, the bee-

eater was thrown off, but it overtook the bustard by flying
and settled again. It was thrown off when the bustard actually took flight."

The famous explorer, Sir Samuel W. Baker, has recorded
another example of hitch hiking birds. When near the Blue
Nile, in Abyssinia, he met what he calls "a remarkably curious hunting-party." He writes:

> A number of the common black-and-white storks were hunting
> for grasshoppers and other insects, but mounted upon the back of
> each stork was a large copper-coloured flycatcher, which, perched
> like a rider on his horse, kept a bright look-out for insects, which
> from its elevated position it could easily discover upon the ground.
> I watched them for some time: whenever the storks perceived a
> grasshopper or other winged insect, they chased them on foot, but
> if they missed their game, the flycatchers darted from their backs
> and flew after the insects like falcons, catching them in their
> beaks, and then returning to their steeds to look out for another
> opportunity.

Baker's "large copper-coloured flycatcher" is probably the
carmine bee-eater, already referred to. Neumann refers to
it again in the following passage. He says it habitually rides
about "on the back of the large crested bustard or 'pauw'
which is common about the north-eastern extremity of
Bassu. It sits far back on the rump of its mount, as a boy
rides a donkey. The pauw does not seem to resent this liberty, but stalks majestically along, while its brilliantly-clad
little jockey keeps a look-out, sitting sideways, and now and
again flies up after an insect it has espied, returning again
after the chase."

Neumann says he has also seen the bee-eater sitting on
the backs of goats, sheep and antelopes, but the bustard appears to be its favorite mount.

These examples of birds riding on birds are of special interest in view of what is to be said later on alleged hitch hike flying.

Short pick-a-back flights are sometimes made by birds when fighting. Those pugnacious flying atoms, the humming-birds, occasionally take temporary flights on the backs of birds invading their nesting territories, while punishing them with their needle-like beaks. According to W. H. Hudson, the Argentine chope can sometimes be seen "pouncing down and fastening itself on the victim's back, where it holds its place till the obnoxious bird has left its territory." (Quoted by McAtee.)

But the bird most addicted to belligerent pick-a-back flying is that indomitable little fighter, the kingbird. So fierce are some of these birds in defense of their territory that nothing that flies can daunt them. In addition to well-authenticated examples of hawks, falcons, eagles and other birds many times the kingbird's size being chased off, an aeroplane has been attacked by one of these birds (see p. 56).

Several instances of kingbird pick-a-back fighting are cited by Bent. Lincoln, for example, says: "I watched a kingbird attack a hawk and saw it alight on the back of the larger bird, to be carried forty to fifty yards before again taking flight."

The kingbird is not the only member of the well-named *Tyrannidæ* to pick-a-back a trespasser off the premises. Florence M. Bailey, writing of the scissor-tailed flycatcher, says she saw one "in pursuit of an innocent caracara who was accidentally passing through the neighborhood. The slow, ungainly caracara was no match for the swift-winged flycatcher, and with a dash Milvulus pounced down upon him

and actually rode the hawk till they were out of sight."
Brandt says these flycatchers "often perch on the caracara's
back for a mile and leave a wake of pulled feathers."

Another form of hitch hike travelling has been questioned in the past, but is now generally accepted. This
concerns the carrying of young by the parent birds. Such
carrying is done in several ways. For example, the young
of the chachalaca have been observed clinging to their
mother's legs while she flew them to the ground from the
nest in a tree. This, of course, is not true pick-a-back flying.
But tree-nesting duck have occasionally been seen to take
their young on their back and fly down with them to earth,
although this is unusual.

The most famous instance of pick-a-back flying with
young occurs with the woodcock. In the past this has been
disputed by many ornithologists, but it has now been witnessed a sufficient number of times by reliable observers to
put it beyond serious question. Bent (1927) quotes several
instances for the American species. Probably the greatest
British authority on woodcock, J. W. Seigne, says he has
seen the parent woodcock fly with young on several occasions when there was no possibility of mistake. The usual
method appears to be between the legs or thighs, but true
pick-a-back flying has been observed as well. In the Woodcock Enquiry, 1934-35, organized by the British Trust for
Ornithology, seven instances were reported where young
were flown on the back of parent birds. According to one
report two young birds were carried in succession across a
river in this manner.

Service reports seeing a lapwing fly over his head "holding betwixt its legs, pressed up against its abdomen, with
its tail at the same time much depressed, what I have every
confidence in saying was a young one. The bird alighted in

the adjoining field, and I marked the spot, and, on running up, found a young one. Many years before I had seen a similar incident, and the belief is very general amongst country folks that lapwings will when any danger threatens, remove their young to safe spots by carrying them."

In his report on the Woodcock Enquiry, Alexander says "correspondents have quoted cases where swans, partridges and grouse, as well as woodcock, have been seen flying with young on their backs." McAtee informs me that mergansers have also been seen carrying young on their back.

There are a number of instances of young birds alighting on the back of a parent while learning to fly. The most famous example of this form of pick-a-back flying occurs with eagles. There is a reference to it in Deuteronomy (ch. 32, v. 11): "As an eagle stirreth up her nest, fluttereth over her young, spreadeth abroad her wings, taketh them, beareth them on her wings. . . ."

As a comment upon this ancient scripture here is a modern account of an incident which was witnessed by F. E. Shuman, a student of Dr. Loye Miller (Bent, 1937), and is quoted by him:

> The mother [golden eagle] started from the nest in the crags, and roughly handling the young one, she allowed him to drop, I should say, about ninety feet, then she would swoop down under him, wings spread, and he would alight on her back. She would soar to the top of the range with him and repeat the process. One time she waited perhaps fifteen minutes between flights. I should say the farthest she let him fall was a hundred and fifty feet. My father and I watched this, spellbound, for over an hour.

A somewhat similar account is quoted by Thomas.

Another instance concerns trumpeter swans. A cygnet was seen flying a few feet above a mature bird, presumably one of its parents. Every now and again the cygnet dropped lightly on its parent's back and, when rested, it took off again, only to rest again in the same way later. (Bent, 1925.)

A variation of this form of pick-a-back flying has been reported by George H. Mackay, a sportsman of high repute, who said he was once out shooting with a companion who fired at, and slightly wounded, a young sandhill crane. After a struggle the bird managed to become air-borne, but its flight was very weak. The parent bird was then seen deliberately to place itself underneath its young, which rested its feet on her back, both birds continuing to flap their wings. A report in *Forest and Stream* recounted a somewhat similar incident in which a wounded Canada goose was alleged to have been helped in flight by another goose (quoted by McAtee).

All these alleged instances of pick-a-back transport are more or less orthodox, however rarely they may have been observed. It is the alleged carrying of small birds by larger, of totally different species, especially during migration flights, that is the controversial question. In the nature of the case numerous exact observations cannot be expected. I think this should be stressed, because even if occasional pick-a-back flying were established as an indisputable fact, I cannot imagine that it would be observed very often. It is only comparatively recently that the carrying of young woodcock in any form was generally admitted, and still more recently was it acknowledged that occasionally the young are carried on the back of the parent. And it is this latter position, the rarest and most difficult to observe of all, which is adopted by the alleged pick-a-backing birds.

The folk-lore of several nations has references to the pick-a-back flying of various birds. It is easy to dismiss such stories as unsubstantiated legend, but anyone who remembers the instances where such legends were right and the science of past days wrong (e.g. the celestial origin of meteorites) will be cautious in summarily dismissing a belief which has been held independently by widely-separated peoples. Although not strictly relevant, it may be remembered that the widespread European tradition of the sovereignty of the wren over all other birds has a pick-a-back basis.

All the birds are said to have met together to decide which should be king. It was agreed that the bird which flew highest should be awarded the crown. The eagle outsoared all its competitors, but after it had proclaimed itself king, the wren, which had secreted itself among the eagle's feathers, flew above it and said: "Birds, look up and behold your king!" (Dyer.)

Several North American Indian tribes believe in the pick-a-back story. The Cree Indians assert that a small brown passerine bird, when migrating, travels on the back of a Canada goose. The Indians assert that they have frequently seen a small bird flying away from geese which have been fired at on the wing.

John Rae, the explorer, says: "An intelligent, truthful and educated Indian, named George Rivers, who was frequently my shooting companion for some years, assured me that he had witnessed this, and I believe I once saw it occur."

Indians living over a thousand miles apart tell the same story. Egyptians had a belief that small birds cross from Europe to Africa on the backs of storks and cranes. Similarly, there is an old Kalmuck saying: "Every crane flying south carries a corncrake on its back." A pretty, if preposterous, embellishment to this story is that, at migration times, cranes circle low over the fields uttering peculiar cries, whereupon their "passengers" fly up and take their seats for the trans-Mediterranean trip. . . .

Is there a germ of truth in these old legends? Von Heuglin, authority on African birds, thought so. He wrote:

> Let others laugh; they know nothing about it. I do not laugh, for the thing is well known to me. I should have made mention of it in my work if I had had any personal proof to justify it. I consider the case probable, though I cannot give any warrant for it. [Quoted by Claypole.]

Before dealing with the evidence proper, I must mention the canary that was carried twenty miles on the back of a homing pigeon! This happened in this country in 1939, and was a stunt used to help to popularize a national wildlife programme. A tiny cock-pit was fixed to the back of the pigeon, and in this the canary nestled during its twenty-mile flight (see plate 61).

It is well-known that small birds sometimes become very tired during their migration flights. It is not at all unusual for them to rest, utterly exhausted, on the decks and rigging of ships which they meet on their way. (Lowery.) E. C. Stuart Baker says that at least two hundred grey quail fell on to a ship on which he was travelling. They "just dropped on to the deck so exhausted that they allowed themselves to be picked up."

Cherry Kearton, in a broadcast talk in England (reprinted in *The Listener,* May 29, 1935), gave a vivid example of how fatigue overcomes small birds when flying against a strong head-wind. They alighted on the ship on which Kearton was travelling, but were frightened away. They tried to fly back. . . .

> But, alas! the wind was too strong for the weak little things, and it was heartbreaking to see their efforts to get back, and I

really felt like jumping overboard to help them. For ten minutes the struggle went on, sometimes the little fellows making headway and almost getting to the ship, only to be swept back by an extra cruel gust of wind. I watched the little victims feebly fluttering, always sinking lower and lower in the air, until they finally dropped exhausted into the cold sea and disappeared for ever.

With that picture in mind, is it so incredible that a small bird, utterly exhausted and about to fall into the sea, should, if a large fellow-migrant were to fly alongside at that moment, alight on its back and possibly cling there until land was reached? As McAtee says: "Knowing that tired migrants alight upon moving vessels, we should not doubt too strongly that they sometimes avail themselves of transport by large birds."

Several interesting records have come from the islands of the Eastern Mediterranean, whose inhabitants are, of course, in an advantageous position to observe migrants flying from Europe to Africa and *vice versa*. Writing in the *New York Evening Post* for November 20, 1880, a correspondent says:

> In the autumn of 1878, on the Island of Crete, the village priest called my attention to the twittering and singing of small birds distinctly heard when a flock of cranes passed on their southward journey. Upon discharge of a gun, three small birds were seen to rise from the flock and soon disappeared among the cranes.

Guillemard had a somewhat similar experience while sailing in the Aegean. A flock of cranes passed over the ship and someone on board shot one. A number of little birds then appeared in the air about the cranes. Van-Lennep also refers to the notes of small birds coming from the backs of larger birds.

A correspondent (E. Hagland) of Ingersoll's had a similar experience in Canada. He wrote:

One fall a flock of cranes passed over me flying very low, and apart from their squawking, I could distinctly hear the twittering of small birds, sparrows of some kind. The chirruping grew louder as the cranes drew towards me, and grew fainter as they drew away; and as the cranes were the only birds in sight I concluded that little birds were taking a free ride to the south.

After I had mentioned some of these incidents in an article, Squadron-Leader W. E. Williamson of the R.A.F. wrote that when he was in Cyprus he found it was a common belief among the villagers that the cranes carried small birds on their backs when migrating. He himself often saw flights of cranes at close range, but never saw any small birds among them. This is hardly surprising if the small birds were on the cranes' backs. He adds that when cranes passed near him after dark, he frequently heard the calls of small birds apparently coming from among the cranes.

Once, when in the Kurdish hills with three other birdwatchers after dark, he noted that "a further flight of cranes came over . . and from among them we all distinctly heard the twittering and chirruping of small birds. The moon had just come over the hills to the east, and some of us were able to get the cranes silhouetted against the moon as they crossed its disc in line, but, even with glasses, none of us could see any small birds. Through that night more flights of cranes passed, calling as they flew, and several times we heard small birds calling with the cranes."

It is difficult to resist the conclusion that some small birds were flying on the backs of these cranes; had they been merely flying *among* the cranes, they would surely have been seen through the glasses.

Further evidence for pick-a-back flying on migration is sometimes provided when migrating birds are shot. A St. Louis journal, in November, 1936, alleged that a humming-bird was found in the feathers of a Canada goose which had been shot at Williams Lake, B.C. While I agree with McAtee that this record may only be an example of "newspaper science," I would point out that, as mentioned on page 193, humming-birds are known to pick-a-back while fighting, and that because an incident is reported in a newspaper it is not necessarily untrue; further, in a subject such as this, where it is, in the nature of the case, so difficult to reach the truth, no scrap of evidence should be dismissed without a hearing. Boulenger says the London Zoo once received a letter from a man in Manitoba saying that he had shot a goose and had found two ruby-throated humming-birds clinging to it. As mentioned on page 197, many North American Indians affirm that small birds have been seen to fly away from geese which have been shot.

Perhaps the most oft-quoted instance of alleged pick-a-back flying was published in *The Zoologist* for February, 1882. The note was sent by T. H. Nelson, who contributed a number of other ornithological notes to *The Zoologist*, and was the author, with W. Eagle Clarke and F. Boyes, of *The Birds of Yorkshire* (1907). The note reads as follows:

The following fact was related to me by Mr. Wilson, the foreman on the South Gare Breakwater, at the mouth of the Tees: I will give the story in my informant's own words, as nearly as possible. Wilson said: "I was at the end of the Gare on the morning of the 16th of October [the day named, the 16th October, 1879, was fine and cold, wind northerly; two days before, the 14th, was the last of the north-east storm which brought the remarkable flight of skuas] and saw a "woodcock" owl (short-eared owl) come flopping across the sea. As it got nearer I saw something sitting between its shoulders and wondered what it could be.

"The owl came and lit on the gearing within ten yards of where I was standing, and, directly it came down, a little bird dropped off its back and flew along the Gare. I signalled for a gun, but the owl saw me move and flew off across the river. We followed the little bird and caught it, and I sent it to Mussell to be made into a feather for my daughter's hat."

The little bird was a golden-crested wren. I have asked Mussell about this affair, and he tells me Wilson gave him exactly the same version as above, and that he has heard him tell the story several times since without the least variation. Wilson could have had no inducement in telling me other than the truth, and I have every reason to believe that what I have written is correct. It does not necessarily follow that the goldcrest came the whole way across the North Sea on the back of the owl; but I think it is quite possible that, feeling tired on the way, it might have availed itself of the assistance of its *compagnon de voyage,* and so be carried to shore. Wilson further told me he had seen another wren on an owl's back about a fortnight after he saw the first one.

Readers will judge this narrative for themselves, but to me it bears the mark of truth.

The incident has been used to try to prove that gold-crests habitually migrate across the North Sea on the backs of owls. But the fact that such unwarranted inference has been based on this slender evidence should not be allowed to rob the incident of the value it has as evidence for pick-a-back flying.

One criticism of this record is that the only known migration of the goldcrest is within the British Isles, and the bird referred to in the report would, therefore, have no occasion to cross the North Sea. But this applies only to the British species: the Continental goldcrest, which differs very slightly from the British, is known to migrate from Scandinavia to the East Coast of Britain.

After I had published an article on this subject in *The Field,* Lloyd wrote a letter, part of which follows, in confirmation of Nelson's statement. Lloyd says:

Mr. Lane's account of goldcrests arriving on the backs of short-eared owls is not surprising to anyone who has seen these birds nearing the coast after a gale. It was my good fortune to see hundreds flitting, scattered between the waves and just above the surface of the water, between one and two miles off the mouth of the Wear early in November, 1899.

A heavy ground swell was running and there was a strong favourable wind towards the shore. Numbers of golden-crested wrens pitched on our tug as it went out to sea. All died when picked up. The time was about 11 a.m.

The case of pick-a-back riding recorded in *The Zoologist* of 1882, quoted by Mr. Lane, occurred after a north-east gale, when migrants would arrive in a similar exhausted state, and an owl flying to leeward of a tired goldcrest would be like a spar to a drowning man, and it would be luck they had the same destination. In a gale and when exhausted, they would be much more likely to come together near the surface of the water than they would in fair weather and flying at variable altitudes.

Nelson is not the only writer who has reported an owl and a goldcrest flying pick-a-back. Haig-Thomas writes:

I know of a sportsman who was waiting for ducks to flight over the marshes in the early morning. As none came he grew bored and, after shooting several gulls and a tern, he shot a short-eared owl that had migrated across the North Sea to hunt rats and mice among the marshes on the east coast. On going to see what he had shot he found a golden-crested wren clinging to the feathers on its back.

I wrote to Haig-Thomas on reading this, and in his reply he said:

I know of three cases of birds carrying smaller ones. One a golden eagle carrying an unknown small bird, two of owls carrying golden-crested wrens. The keeper who told me about the golden eagle and one man who told me about the owl and a wren are dead. The other case was told me in 1928 by Frank Cringle, a gunner of Wells. I don't know where he got his information from. I think it was by hearsay only; the other two cases had been seen personally by those that told it to me.

Henry Williamson records a personal observation from the East Coast of a firecrest, a close relative of the goldcrest, being found on an owl's back. He writes:

> I saw a woodcock [short-eared] owl fall shot among the marram grasses near the edge of the sea, where small sand-dunes are being made by the east winds. On the owl's back, clinging to its feathers, was a tiny bird. It was a fire-crested wren, exhausted by the long journey from Scandinavia. . . . Warmed in the hand, the mite with the vermilion streak on its head opened its eyes and flitted away over the marshes.

Williamson says on another occasion an East Anglian wildfowler told him he had refrained from shooting an owl for fear of hurting "the li'l old totty thing ridin' 'long with him." It would appear from this remark that it is a common tradition among the East Anglian wildfowlers that these tiny birds, exhausted by their long flight across the North Sea, occasionally complete the last stage of their long journey on the broad backs of their fellow migrants, the short-eared owls.

Such, then, is the best of the evidence I have collected on the subject of pick-a-back flying. For the most disputed form of pick-a-backing, it is, admittedly, not strong. McAtee says: "Manifestly pick-a-backing cannot be a major factor in migration. Yet, as to its simple occurrence, there can hardly be disbelief." I certainly think it is a subject which should not be summarily dismissed, and if this chapter arouses others to endeavour to obtain additional evidence, then it will not have been written in vain.

11

Some Experiments with Animals

A SCIENTIST once gave a lecture in which he described a number of ingenious experiments which had recently been carried out. After the lecture a lady came up to him and said: "Excuse me, but what *use* are all these experiments?"

The scientist hesitated, gave a lame apologia for his life's work, and then said with a smile: "We carry out these experiments because we find them amusing." When Faraday was once asked a similar question, he blandly replied: "Madam, will you tell me the use of a newborn child?" A similar answer is attributed to Benjamin Franklin when asked the value of balloon ascents. (*Nature*, February 16 and March 9, 1946.)

Actually, of course, neither scientist was doing himself justice. Experiments with animals are of considerable value from various aspects. Practically they sometimes help the farmer, the apiarist, the horse or dog breeder to get better results from their stock. Lessons learnt from animals in the laboratory are frequently of value in the treatment of human beings in hospital.

But it is not always possible to point out the practical value of a given experiment. It may, in fact, not become apparent until years after, when additional information becomes available and the earlier experiment is then found to fit into place like the missing piece of a jig-saw puzzle.

But natural history would never have got far if its students had been concerned only, or even primarily, with the practical application of the secrets they wrested from Nature. Experiments with animals enable us to probe a little deeper into their minds and natures, and this is sufficient justification for the work described in the following pages.

It should, however, be pointed out that many experiments create artificial conditions and these may, in consequence, affect the behaviour of the animals being investigated. While this point should be mentioned, I think it is true to say that the experiments mentioned in this chapter revealed aspects of the physical and mental attributes of animals which it would have been difficult, if not impossible, to have discovered in any other way.

In its natural state situations rarely occur which compel an animal to use its senses to the utmost. But in the laboratory and under controlled conditions in the field, it is possible to devise experiments which put to the ultimate test the powers of an animal and thus to find out the maximum abilities of which it is capable. And it is often only by testing until the animal *must* fail that its maximum abilities can be determined.

The following accounts of some experiments with dogs will illustrate this point. Most people know that a dog has a very keen sense of smell. But just how keen is it? It is only by such tests as those to be described that an adequate answer can be given.

In the first place it should be remembered that a dog is, above all, a smelling animal. Hearing comes next and sight last. In man, with whom sight is the prime faculty, a small space only in the upper part of the nose is used for smelling, but the olfactory membranes in a dog's nose cover a much more extensive area.

Buytendijk carried out several experiments with a police

dog (presumably an Alsatian) named Albert. The dog was first tested in a laboratory, where it was found that it could recognize sulphuric acid in a dilution of 1:10,000,000. Then Albert was taken into the open-air and the following experiment was arranged.

Each of six people standing together threw a stone on to some gravel. Albert had been allowed to smell one of the men's hands and he was now ordered to retrieve the stone thrown by that man from among the six which lay among the gravel. Buytendijk says:

> He did this, sniffing round, standing still at every stone that had been thrown, until he came to the right one and brought it back. In this test it was very remarkable to see how the behaviour of the dog altered the moment he found the right stone. While sniffing leisurely at each spot in turn he suddenly pricked up his ears; with quick movements of nose and mouth he freed the stone from its surroundings and speedily brought it back. The behaviour of the animal was entirely different when, either influenced by the trainer or for some other reason, he took up a stone that was not the right one. Then his reaction was uncertain, almost shy.

Remarkable as was this performance, other dogs have performed similar feats which are no less surprising. Romanes has recorded the wonderful ability of a favourite setter bitch to follow his track. Romanes got eleven men to walk behind him in Indian file. Each man following was careful to place his feet in the footprints of his predecessor.

When the procession had walked two hundred yards Romanes turned to the right and five men followed him. The rest turned to the left. The two parties continued walking for a considerable distance and then concealed themselves. The setter was then put on the common track before it diverged. She found Romanes without difficulty.

It had previously been ascertained that the setter would not follow the scent of any other man in the party, except, first, Romanes and then the gamekeeper. Romanes said: "Yet my footprints in the common track were overlaid by eleven others, and in the track to the right by five others. Moreover, as it was the gamekeeper who brought up the rear, and as in the absence of my trail she would always follow his, the fact of his scent being, so to speak, uppermost in the series, was shown in no way to disconcert the animal following another familiar scent lowermost in the series."

But perhaps the most exacting test of the olfactory powers of a dog were applied, not in a scientific test, but in real life. This occurred when an Alsatian attached to the Cairo police was called in to follow the four-and-a-half-days-old track of a donkey across rocky ground. The dog succeeded and stopped, barking, outside the house where the donkey was kept! (Arundel.)

It may be thought that the dog has no rivals in Nature in its power to detect the most faint smells. Yet some species of moths possess an almost incredibly keen sense of smell during the mating season. At periods during this season the female sends out a call-sign to all males in the neighborhood that she is ready for mating. She does this by extruding her scent organ.

The human nose is capable of detecting three one hundred million parts of one grain of musk, yet when an Emperor moth is "assembling," as entomologists call this mating rite of the moths, no human nose can detect the faintest whiff of scent. Yet male moths will be aroused by it from afar and will fly to the female from a distance of half a mile.

Dr. T. C. Schneirla comments: "Strictly speaking, I

doubt such distances as a matter of direct flight, of 'flying to.' . . . Some of the orientation experiments suggest a possible way: gradual arousal and wandering first, gradual approach of some individuals after many circuitous movements, finally, *direct* flight through a given distance."

Riley obtained an even more striking result in an experiment with the moth of the Japanese ailanthus silkworm. He confined a female in a small wicker cage outdoors and then released a male, with a silk thread tied round its abdomen for identification, at least a mile and a half away. The next morning he found the marked male at the cage.

Some entomologists have wondered if instances such as this can be explained only on the assumption of some form of telepathic communication which exists between a calling female and the responding male moths. But as calling females confined in air-tight containers are unable to summon a single male, this explanation would appear to be untenable.

It is not surprising, however, that such an explanation has been put forward. Think what "calling" by scent entails. It means that scent particles, which even in high concentration are not detectable at all by the human nose, sometimes impregnate half a cubic mile of air. (In the instance cited by Riley it would be 13½ cubic miles.) Wind conditions obviously play a large part in the dispersal of the scent, but my figures appear to be accurate in order of magnitude. Put in another way, it means that a fraction of a grain of moth scent is dispersed in some half a million tons of air and yet can be detected by the male moth! See also Kettlewell.

Several experiments have been made to determine how the scent of ants is used as a means of identification. One of the earliest experiments was made by Sir John Lubbock.

He took a number of ants from each of two nests and made them so drunk that they became insensible. He then marked them with differently coloured spots and put them all on a table near some ants from one of the nests. The table was surrounded by water to prevent the ants from wandering away.

At first the sober ants were rather puzzled by the strange behaviour of their intoxicated companions. After examining them, however, they apparently decided on their course of action. The intoxicated ants from their own nest were carried home, while the ants from the other nest were dumped into the water! Apparently scent alone, even when mingled with the vapors of alcoholic excesses, was sufficient for one ant to recognize another from the same nest.

More recent experiments have confirmed Lubbock's general findings. It appears that each nest of ants has a peculiar odor which is the basis of the distinction between friends and foes. Bethe took an ant from one nest, dipped it first in weak alcohol and then in water and then in the juice obtained by crushing together the bodies of a number of ants of another species. He found that such an ant, if replaced in its own nest, would be killed by its erstwhile companions, but it could be safely introduced into the nest of the ants whose odor it now bore, even although its appearance was quite different from that of the other ants in the nest. But when the artificial odor wore off and the ant's own scent became apparent, there was liable to be trouble!

The sense of hearing in various animals has also been the subject of a number of experiments. Although, as mentioned above, the dog is not primarily a hearing animal, tests have shown that its ears are remarkably acute. Engelmann found that a faint sound inaudible to human ears beyond a distance of thirteen feet could be heard by a dog at a distance of seventy-six feet.

Both Engelmann and Katz carried out a number of experiments on the ability of dogs to localize sound. The dog used in most of the experiments was an Alsatian bitch. It was first trained to respond to an electric buzzer placed behind a small board. The buzzer was sounded for a third of a second only. When fully trained to respond to this sound, the dog was placed in the centre of a circle twenty feet in diameter. Around the circumference of the circle, at regular intervals, were placed sixty small boards, behind any one of which the buzzer could be sounded. In almost every test which was made the dog went to the board from which the sound had come. A buzz coming from a board behind the dog was localized as correctly as from one in front. Every precaution was taken to see that the dog got no help from any source other than its ears.

That these results are typical of the performance of a first-class dog was confirmed by Keller and Brückner. They placed a dog within a circle twenty-five feet in diameter, around the circumference of which were sixty-four boards. Here again the dog was able to go immediately to the board from which the buzzer had momentarily sounded.

These workers subjected the dog they were using to another test. Two boards were placed six inches apart and the buzzer could be sounded from behind either of them. The dog had to go to the board from behind which the buzzer had sounded. Even when the boards were sixteen feet away from the dog it could distinguish between them.

Katz believes that the ability to localize sound in this way is owing to the dog's appreciation of the difference in the time which it takes the sound to reach its ears. (Keller and Brückner, however, doubt this explanation.) When one ear of a dog was stopped it could no longer localize sound correctly. In fact, it did better with both ears stopped than with one stopped and one not.

Referring to the second test by Keller and Brückner, Katz says: "Assuming the theory of difference in time to be correct, we find here a difference of not more than 0.0000045 seconds. . . . The calculation is made on the basis of the distance between the dog's ears and the distance between the source of sound and the two ears."

Katz continued these experiments with several modifications. First he provided the sources of sound from various heights. He found that the dog could still readily distinguish these when they were up to the height of its head, but above that it became unreliable. This result is interesting when seen against the dog's normal background. The whole sense world of a dog—sight, sound and smell—is normally bounded within three feet of the ground. Beyond that, except for signals from its master and rare excursions, it never goes. And when test sounds are emitted from even the fringe of that larger world, the dog's otherwise wonderfully acute sense of hearing begins to break down.

Katz carried out a further hearing test with his dog. This was to determine whether it could distinguish three sources of sound lying in the same plane, one behind the other. But here the dog failed completely. It ran in the right direction, but stopped at the first sound board, irrespective of whether the sound was actually coming from that, or from either of the two beyond it.

Katz later experimented with hearing in cats. He found that they would not respond to the buzzer which had been used for the experiments with dogs, but they would to a noise which resembled a running mouse! One cat could distinguish between two sounds twenty inches apart from a distance of sixty feet. This gives a time difference of 0.0000028 seconds, which is considerably less than that for the dog Keller and Brückner tested.

When cats were tested for their ability to localize sound at varying heights and distances, unlike dogs, they were found to be very efficient. Their manner was also noticeably different from that of dogs. Instead of dashing off immediately the sound was given, as would a dog, the cat waited, moving its ears, thus making sure exactly where to go before it set off.

The difference, both in manner and ability, between the dogs and cats used in these experiments, is no doubt related to their habits of capturing prey. The dog, with its wolf ancestry, runs toward a likely noise and then, relying on its eyes, chases whatever may come into view. The cat captures its prey by quietly manœuvering into the most favorable position and then making one sudden leap, often in the dark. In such hunting an exact sense of location is indispensable. Similarly, a cat seeks its prey in trees as well as on the ground, and therefore has much greater need to be able to localize sounds at varying heights than has a dog, which almost invariably seeks its prey on the ground.

Experiments on the power of sight in animals have revealed that by far the most efficient eyes are possessed by birds. The avian eyeball is of relatively huge size. Normally only the comparatively small cornea shows in the circular lid-opening, but the entire eye is so big that there is literally often barely enough room in the bird's head for the two of them. Hawks and owls, with bodies only a small fraction the size of a man's, have eyeballs which are as large or even larger.

A bird's two eyes often weigh more than its brain. The largest of all, and also the largest eye of any land vertebrate, is that of the ostrich, which measures two inches in diameter. Two measured by Crile weighed 95.26 grams, which was over twice the weight of the ostrich's brain, 42.11 grams. Crile lists a number of other birds whose eyes weigh more than their brains. Several of these follow, all weights in grams.

Bird					Eyes	Brain
Greater bustard	–	–	–	–	52.45	15.63
Fish eagle	–	–	–	–	22.28	12.93
Tawny eagle	–	–	–	–	32.84	14.09
White Orpington fowl	–	–	–		6.39	3.55
Red-tailed hawk	–	–	–	–	21.22	10.025

Thomas Shastid, an ophthalmologist who has specially studied the eyes of animals, says of birds' eyes:

> These are the finest and most remarkable of all the eyes of earth, being often both telescopic and microscopic. In birds the visual acuteness is almost incredible, in some instances a hundred times as great as that in men. A bit of grain that human eyes can barely see at a distance of one yard, a bird can see distinctly at a distance of a hundred yards.

This appears to be an exaggeration. Smith says "the visual acuity of the kestrel's eye is approximately eight times that of the human eye," and "the sensitivity of the owl's eye in conditions of low light intensity has been shown experimentally to be about ten times that of the human eye under similar conditions."

Schmid carried out several experiments with a peregrine falcon and a tiercel (as the male is called) in Germany, in 1934. The object of the tests was to see from what distance the peregrines would fly to the lure, the name given to the small bunch of feathers which, when swung on the end of a cord by the falconer, recalls the trained bird to him. In Schmid's experiments the lure consisted of two rooks' wings bound together. The distances in the experiment were measured by a range-finder and checked later.

In one test the lure was swung when the falcon was 4,300 feet away and she flew to it immediately. The tiercel did even better. On a day when visibility was far from ideal, he responded to the swung lure from a distance of 5,400 feet, or over a mile. At a shorter distance than this from the lure, one of the experimenters was unable to recognize it through 6X binoculars. Schmid ends his account of these tests by saying: "Personally, I have no doubt whatever that falcons are able to see the feather lure or the living quarry at still greater distances."

The experiments dealt with so far have been concerned largely with testing the physical senses of various animals. But by far the largest number of experiments have been concerned with their intelligence and general psychology. It would be impossible within the short compass of this chapter to cover all of this ground, however cursorily, and I shall therefore deal only with some experiments which are interesting either in themselves, or for the information they yield concerning the character of the animal mind.

Probably no other insect, and few other animals, has been the subject of so many experiments as the bee. Foremost among those who have studied the ways of this wonderful little creature is Professor Karl von Frisch (1923, 1937 and 1951), of the University of Munich. He was particularly interested in the ways in which bees communicated to one another information concerning the honey yields to be found in the various flowers they visited, in other words, "the language of bees." Von Frisch determined to unravel their secret.

On a small table in the meadow of the Munich Botanical Gardens he placed a small blue card, and on this a watch-glass containing honey. The bees, finding this rich source of food, came to it regularly. After a while the original blue card was removed and, on either side of the spot where it had been, two new cards were placed, one red, the other blue. When the bees next came to the table they flew direct to the blue card.

This experiment showed that the bees could distinguish colors. But it did not prove that they necessarily possessed color-sense. A color-blind man, for example, can also distinguish blue from red, because the blue appears as a lighter shade of the all-prevailing grey than does the red.

Von Frisch's next experiment, therefore, was to see if bees possessed a real color-sense. When the blue card was placed among others of all shades of grey, it was invariably picked out by the bees, even when there was no honey on it. It was found that bees could equally be trained to come to orange, yellow, green, violet and purple cards. But when all the differently-colored cards were placed on the table together, the bees became somewhat confused. Evidently their color-vision is not so good as ours. But, unlike us, they can see ultra-violet rays. Bees also appear to be color-blind to red. This is particularly interesting, because there are very few red flowers which are fertilized by bees.

Von Frisch (1919) also tested the bees' sense of smell. On his experimental table he placed a number of small cardboard boxes, each with a hole in front. In only one box was honey placed and this box was marked with a distinctive odor. The bees found the box containing the honey. Then its position among the other boxes was changed but, guided by the scent, the returning bees made unerringly for it.

Later all the boxes were replaced by new ones. One of them was marked with the distinctive odor the bees had learned to associate with food, but this time no honey was put inside. The bees flew to the new boxes, flew around the holes but entered only the one that was scented. In other experiments all the boxes were provided with different scents, but the bees could pick out the scent that meant food from forty other scents. Even when the feeding scent was much diluted it could still be recognized. From all these experiments von Frisch concluded that the sense of smell in bees is about the same as for human beings.

In other experiments bees were trained to come to a yellow box with a rose scent. After a while another series of boxes was put out. The one yellow box had no scent, one white box had the rose scent and all the other boxes were white with no scent. What would the bees do now? They flew towards the yellow box, but, when they got near it, they veered away and went to the one white box with the rose scent. It would appear from the results of this experiment that bees recognise flowers from a distance by their colors, but when they get near them scent is the predominant quality.

During his experiments von Frisch noticed that when he placed honey on his experimental table, it was many hours, sometimes days, before the first bee discovered the food. But as soon as one bee found it, many, sometimes hundreds, came within a short time. Moreover, they all came from the same hive as the original discoverer. Von Frisch was now near to finding out the secret of the language of the bees.

He constructed special observation hives with glass windows and a special arrangement of the wax combs so that he could watch what went on in the interior of the hive.

He then arranged a special system of marking, with colored spots of paint, the bees which came to his experimental table. By this system he could number up to 599 bees and could distinguish them even when they were in flight. He thus describes what happened to the first bee that came to the table and was marked. (Von Frisch, 1937.)

First, it delivers the honey or sugar water, found and sucked up on our table, to other bees in the hive. Then it begins to dance. On the same spot it turns round and round in a circle with quick, tripping little steps, once to the right, once to the left, very vigorously, often half a minute or a full minute on the same spot. It is not possible to give a good description in mere words. The dance finishes just as suddenly as it began, the bee hurries to the hold of the hive and returns to the feeding-place.

The bees on the wax comb around the dancing bee become greatly excited by the dance; they trip behind the dancer, following all its turning movements. They turn their heads to it and keep their feelers as closely as possible to its body and it is evident that they are highly interested. Suddenly one of the following bees and then another turns away, cleans its wings and antennæ, and leaves the hive. Soon afterward these new bees appear at the food place. After homing, they dance also and the more bees there are dancing in the hive the more appear at the feeding-place. It is clear that the existence of the food is communicated by the dance in the hive.

Von Frisch had discovered the "language" of the bees!

He now wanted to know how the bees that followed the original dancer, knew where to go. They do not follow it directly, as they appeared at the feeding-place independently. Another experiment gave him the clue.

He spread feeding-places round the hive in all directions. A few minutes after the bee which discovered the first feeding-place had returned to the hive and danced its good news, bees appeared and flew out in all directions. They settled on all the feeding-places. Now if the first dancing bee had indicated the particular feeding-place it had come from, the bees would all have flown to that.

Von Frisch concluded that the dance was only a general message that food was available and the bees must fly out and look for it. He adds:

> When there were no dances in the hive, the little glass dishes in the meadow were not visited by any bee for many days. As soon as there were dances in the hive, the dishes in the neighbourhood were all found within the shortest time.
>
> But not only in the neighbourhood! In further experiments I left the feeding-dish, visited by some numbered bees, at a short distance from the hive. And I put some other dishes farther and farther away in the meadow, observing whether they would be found or not. The farther they were the longer time it took till they were found by the bees sent out by the dancer. In the last experiment they were found after four hours in a meadow a full kilometer [about half a mile] from the hive, with hills and woods lying between them. It is clear from a long series of experiments that, after the commencement of the dances, the bees first seek in the neighbourhood, and then go farther away, and finally search the whole flying district.

But still von Frisch was not satisfied. Feeding from glass dishes was not natural for bees: suppose flowers were used in the experiments instead? Into some cyclamen flowers he dropped some sugar-water. Nearby on the ground he placed a bunch of cyclamen and a bunch of phlox. The bees came as usual, but they were interested only in cyclamen. The experimental flowers were later reversed, the sugar-water now being placed in the phlox and not in the cyclamen. The bees were now interested only in phlox, not only the experimental bunch, but all those growing in the neighbouring gardens. This was particularly interesting because normally bees never visit phlox.

Von Frisch experimented with numbers of other flowers and succeeded in getting bees to respond in the same way with all of them except those without scent. He concluded that the bees, watching the dancing bee when it returns to the hive, hold their antennæ against its body and perceive the specific scent it has picked up from the flowers it has just visited. The other bees then fly out to seek the same scent.

Yet another factor in the bees' language is the special scent organ they possess in a pocket of the cuticle. This is normally closed and then can give out no scent. But if a bee discovers a rich source of food, it extrudes the scent organ. The scent given out has a powerful attraction for bees and tells those following the dancer's directions when they are in the right vicinity, attracting them from a considerable distance. When von Frisch shellacked over the scent organ of the bee which originally found a feeding-place, he found that bees arriving later discovered the food only by chance. Another use for the scent organ is to guide bees to odorless flowers which are yielding honey.

In more recent experiments von Frisch (1951) has shown that in the dance the bees can indicate the approximate distance of the source of food from the hive. Generally this is done by the tempo of the dance, the further the distance the slower the dance. In the same dance they also indicate the direction in which the food lies, using the sun as a point of reference.

Von Frisch told me, when I saw him in 1949 at his bee laboratories at Brunnwinkl in Austria, that he once carried out the following experiment. He got an assistant to put out some food but von Frisch did not know where. When a bee found it it was distinctly marked, and von Frisch carefully watched its dance when it returned to the hive. He read the message it danced to its fellows and then he said: "The food has been placed 320 metres from the hive" and he gave the direction. Checking showed that actually the food had been placed 332 metres from the hive, and in a direction which von Frisch had estimated correctly to within four degrees.

A Kentucky farmer was unfortunate enough to have a practical demonstration of how bees concentrate their efforts where rewards are greatest. He had taken seventy-five pounds of surplus honey and stored it in a garage not far from his hives. A bee discovered this golden hoard and was soon dancing excitedly in the hive. More bees went out and likewise came home dancing. This process continued, so vast was the harvest to be gathered, and in the end all the bees in all the hives were collecting honey from this one store. When the farmer discovered what his bees were doing, nearly the whole of the seventy-five pounds had been transported back to the hives! (Teale.)

Von Frisch's experiments in Germany have been elaborated by the work of Julien Françon in France. In one experiment he placed a saucer with moistened sugar on the grass. When a bee took food from it, he marked it with a yellow spot. Soon reinforcements came and were similarly marked with yellow. When eighteen bees had thus been marked, Françon placed another saucer ten yards away from the first one. Bees came to this and twenty-three were marked with red spots.

By examining closely the two saucers, Françon never found a bee with a yellow spot at the saucer where the red-marked bees were feeding, or *vice versa*. This was the more remarkable as the red-marked bees, to reach their saucer, had to fly directly over the saucer where the yellow-marked bees were feeding. Yet no mistakes were made; each batch of bees went to the saucer they were used to. But when Françon drastically reduced the amount of food

in the yellow-marked bees' saucer, he found the number of
bees coming to it was rapidly reduced and yellow-marked
bees now began appearing at the saucer for the red-marked
bees!

Françon was particularly interested in the way a bee
would return to a given spot and how it could apparently
direct other bees to it. He caught a bee and placed it on
some sugar on a saucer. The bee took its fill of the sugar
and then flew off. But it was a very strange flight. Françon describes it thus:

> Slowly she traces three small circles at no great height from the
> saucer, which lies on a chair, her head constantly turned towards
> the centre, from which she gradually recedes in a regular spiral,
> and, so to speak, backwards. Evidently she is examining with the
> greatest care the place she is leaving, as though to make a correct
> survey of its disposition. Two more larger circles, higher and
> faster, a last curve, and then the rectilinear flight, twenty-five to
> thirty feet above the saucer straight to the east, in the direction
> of the distant apiary of the hamlet.

Françon believed that this survey flight, as he calls it,
preserved in the bee's memory the location of the saucer
with mathematical accuracy. If the bee was prevented from
making the survey flight, it was quite unable to find the
saucer again. Françon found that when, after the bee had
made the survey flight and had flown back to the hive, he
moved the saucer a few feet, the bee would return to the
exact spot where it had been and would seem greatly per-
turbed to find nothing there.

"She cannot believe her own eyes. She persists, flies
close to the ground between the blades of grass, rises,
moves away, verifies her alignments, and comes back, always to the same spot. Close to, the saucer forms a very
conspicuous white patch. The bee ignores it. She recog-

nizes nothing but the mathematical point which is her aim." The bee behaved in the same manner when there was a vertical displacement of the saucer of only eighteen inches.

Opfinger, however, has challenged the importance of the orientation flight and considers the appearance of the feeding-place on the *arrival* of the bee more important in fixing its position on the insect's memory. (Dr. T. C. Schneirla tells me: "This characteristic seems to be an important one. I found it independently in maze work with ants.") This is not the first contradiction I have noticed in the work of investigators of bee behavior. What the solution is I do not know, but undoubtedly there is still much more to be learnt about bees and their complicated ways of life.

Having carried his researches thus far, Françon subjected his bees to some experiments designed to test their peculiar powers to the utmost. He caught a bee and placed it in a cardboard box, the entrance to which was through a narrow tunnel eight inches long. Inside the box at the end of the tunnel was sugar. The imprisoned bee fed on the sugar, crawled down the long exit, made the usual careful survey, this time of the end of the tunnel, and flew back to the hive. Four times the bee, which had been marked with red for identification, returned alone.

On the fifth journey another bee flew back with the marked one. Françon then witnessed a remarkable incident. While the marked bee was still hovering over the tunnel's entrance, the new bee boldly entered it—fully a minute before the one which, it would have been thought, should have shown it the way.

Later Françon blocked the entrance to the tunnel and fixed a paper chimney into the roof. The bees, both the original discoverer and those which followed it, now ignored the blocked tunnel entrance and flew direct to the chimney, down which they clambered to the food inside. He also hid the box and tunnel entirely, by covering it with either sand or by a mass of leaves, stones and bricks. And under these most exacting conditions, utterly unlike anything the bees encountered in their natural state, nearly all the bees which made the survey flights after emerging, and those that followed, found the entrance.

Françon now decided to find out how bees would behave if they were given false information. Using the box and tunnel already described, he allowed a bee, after finding the sugar, to escape, not by the tunnel, but by a crack in the lid. After it had gone he closed up the crack. When, later, other bees came they sought industriously for the now non-existent crack in the lid. They made no attempt to enter by the tunnel. Later the original bee was made to use the tunnel, as in previous experiments, and now the bees that followed it went straight to the tunnel.

All these experiments, carried out by Françon over a period of years, would indicate that bees communicate in a language which can be remarkably precise. How else can we account for the exact knowledge shown by the bees which follow the original discoverer, of the position, and means of access, to stores of food hidden in the most unnatural surroundings?

As H. Eltringham says in the Preface to his translation of Françon's book: "It seems impossible to suggest how such achievements can be attained unless the original bee can, in some way unknown to us, give to the assistant bees the most precise and accurate instructions." (At the same time it must be pointed out that Françon's experiments lacked some essential controls, especially on chemical cues (e.g. scent-gland effects), which lessen the value of his work.)

Well might Françon conclude: "But shall we ever know what goes on behind their domed brows, in the depths of their unmoving eyes?"

An experiment carried out by Möbius, with a pike, shows how radically an animal's behaviour can be modified by experience. He placed a pike in one half of an aquarium and in the other half, separated from it by a glass partition, he placed some minnows. Immediately the pike made a dash for the minnows, but succeeded only in bumping its head against the glass partition. It tried again with the same result, and so on, many times. Glass was something altogether outside the range of the pike's previous experience.

Eventually the pike stopped trying to get the minnows and no longer bumped its head against the glass. Then Möbius removed the partition altogether. What would the pike do now? Although the minnows swam freely round the pike it showed no interest in them! Apparently the painful bump on its head every time it had attempted to eat a minnow had made it thoroughly allergic to them. Triplett carried out a similar experiment with perch, and here again, after a suitable period of conditioning, the perch left the minnows alone when both were allowed to swim about together.

Katz also made use of the glass partition technique in an experiment with a cockerel. A glass partition, with strips of paper stuck on it to make it obvious, was placed partway across a small compound, but a gap was left between the end of the glass and the farther wall of the compound. Some food was placed behind the glass and on the other side was placed the cockerel. To get to the food the cockerel had, of course, to make a detour round the glass.

After the cockerel had become accustomed to going

round the glass to obtain its food, the partition, with the distinguishing strips of paper, was removed. But still the cockerel went the long way round to obtain its food.

A somewhat similar reaction has been observed with rats which have learnt to run a maze. If, when they had learnt it thoroughly, the alleys were either lengthened or shortened, the rats were greatly disturbed. In a shortened alley they often ran headlong into the end walls. Where the alley had been lengthened, they often tried to make a turn at the point were the junctions had previously been, but where now, of course, was solid wall. (Carr and Watson.)

Katz carried out a number of experiments on the feeding habits of domestic fowls, the results of which, apart from the light they throw on the birds' psychology, are also of interest to the chicken-farmer. A heap of grain was placed before a hen which had not been allowed to feed for twenty-four hours. When the hen ate until it stopped before the remaining food without touching it, Katz concluded that it had completely satisfied its hunger.

Katz then experimented with different hens all in approximately the same degree of hunger by using various kinds and heaps of grain. Before one hen he would place a very large heap of grain, before another a slightly smaller heap, and so on, until with the last hen there was not very much more than the hen would need to satisfy its hunger completely. In the next series of experiments the hens would be changed round: the hen which, in the first experiment, was place before a small heap of grain would now be placed before a larger, and so on.

As a result of these experiments Katz found that when a hen was placed before the larger heaps of food, it would eat from thirty-five to fifty-four more grams of food than when placed before the smaller heaps. In other words, irrespective of hunger or the type of food offered, the amount of food the hens consumed increased with the amount presented.

In another experiment Bayer allowed a hen to eat its fill before a large heap of food, and, when it had thus eaten to satiety, the remaining food was removed and then immediately replaced. The hen started to eat again. When it stopped the second time the food was again removed and again immediately replaced as before. Again the hen restarted eating. Some hens recommenced eating under these conditions as many as eight times, and ate 67 per cent. additional to the amount they had consumed when the food was first placed before them.

A fully satisfied hen could also be induced to start feeding again by bringing in a hungry hen and letting it feed on the heap of food the first hen had left. Some hens would eat an additional 60 per cent. in these circumstances and if, later, a third hen was brought in, the first would eat yet again.

Katz tried the experiment of letting hens feed off a hard surface, such as a wooden table, and on a soft surface, such as felt. He found that the hens feeding on the soft surface ate much more, sometimes twice as much, as those feeding off the hard surface. He inferred that pecking on a hard surface causes painful vibrations to a hen's beak and these increase in intensity the longer the hen feeds. It therefore finishes much more quickly than it does when feeding on a surface where the pleasure of eating is unalloyed.

Bayer made an interesting discovery about the pecking of hens when he fed some on grain placed on a putty plate.

He found that the holes made by the beak of a hungry hen were deeper than those made by the beak of a hen that was less hungry.

The Nobles once discovered that a pair of Northern flickers were preparing a nesting-hole in a willow tree in their garden. These birds are practically identical in appearance, except that the male has a small band, or mustache, of black feathers on his breast. (See Plate 72.) The Nobles decided to try the experiment of putting an artificial mustache on the female in their garden. Accordingly she was trapped and some black feathers were gummed on her breast.

On being released, the flicker flew straight to the nesting-hole and alighted on its edge, waiting for the absent male. A few moments later he came. He dropped down beside her and then flew to a nearby branch.

It was some moments before he realized that something was wrong, but when he did, as far as a bird can be said to throw a fit, that male flicker threw one then. He nearly fell off his perch with shock and then, recovering his balance, made a vicious lunge at the female. Completely deceived by her male mustache, he took her for a hated rival brazenly trespassing on his territory.

For over two hours the male chased the female all over the garden. To his vicious pecks the poor bewildered female answered with a soft conciliatory *we-cup.* "Again and again she returned to the hole, only to receive such an onslaught that she lost her balance and fell several feet before catching hold of a branch. After ten minutes of rough treatment she led the male away from the tree and five minutes later he returned alone. Two minutes later the female came back and again *we-cupped* softly as the male drove her out on a limb. . . ." (Noble, 1945.)

Eventually the female was driven into a culvert, where the Nobles rescued her and placed her in a cage until the next day. When she was released minus the offending mustache, and returned to the nesting-hole, the male flew down to greet her. But she had not forgotten the outrageous treatment she had received the previous day. It was now her turn to repay a friendly advance with a vicious peck. The male did not peck back, but sat quietly nearby, giving an occasional subdued *we-cup* until he was eventually restored to favor.

Another experiment with birds, but of a very different kind, was carried out by Marais. This was with the South African yellow weaver-birds. These birds frequent reed beds and make an elaborate nest. The nests are rounded and suspended between upright reeds. Externally they are made of coarse strips of reed blades and internally are lined with soft grasses. The entrance is through a hole in the bottom of the nest. (Roberts.)

Marais wondered what would happen if several generations of weaver-birds were bred without any knowledge of the nests made by their ancestors. Would the young of the last generation, if released in their natural environment, know how to build the intricate nests of their forebears?

Marais took a pair of weaver-birds and isolated them from all nesting material. He took the eggs they laid and hatched them under canaries. When two of the birds from these eggs mated, again without having access to the smallest scrap of nesting material, the eggs were again placed under a canary.

This process continued through four generations of weaver-birds. Then the fifth generation was allowed to nest how they pleased. Without hesitation they made the elaborate nest of their distant ancestors. "No one taught these birds. Four generations of their ancestors had never seen a plaited nest, yet the fifth generation remembered what to do."

It would have been interesting if Marais had been able to isolate not four, but forty generations of weaver-birds and then to have seen how the forty-first pair behaved. I say this because Verlaine, experimenting with canaries descended from countless generations reared in artificial nests, found that when he supplied a pair with suitable materials, they took a long time to make a nest and when it was finished it was shapeless. But nests they made subsequently were constructed more quickly and were of better shape.

Probably no other mammal has been made the subject of so many experiments as the rat. In fact, a whole book has been written about experiments with rats—*The Rat in Laboratory Investigation* (1942), by John Q. Griffith, junior, and Edmond J. Farris, containing over five hundred pages, illustrations and numerous references. From the mass of available material I have selected an experiment of Mowrer's, one of the most ingenious tests of animal psychology that could be devised, and with an unexpected result.

Mowrer installed a feeding apparatus in a cage whereby a lever had to be pressed in order that a pellet of food should drop from a slot. Lever and slot were side by side. Three rats were placed one at a time in the cage and soon learnt the technique of obtaining food.

Then Mowrer had the lever put on the side of the cage opposite to the food slot. This arrangement made it necessary for a rat to run from one end of the cage to the other for every piece of food. Again the three rats learned, separately, how to obtain their food.

Now the three rats were placed in the cage together. For

the first day the three rats hovered near the food slot. Each rat was unwilling to cross the cage and press the lever, as invariably it lost the food it had earned. The rats behaved on the second day in a similar manner. Occasionally one or other would press the lever, but the food thus produced was immediately eaten by the two rats hovering near the food slot.

On the third day the rats were so ravenous that they tried to chew the steel food slot. On the fourth day one rat pressed the lever three times in quick succession and dashed across the cage just in time to get the last piece of food. The same rat then returned to the lever and pressed it a number of times, thus releasing several food pellets. By dashing back to the food slot it was able to eat one or two pellets that the other two rats had not yet had time to consume. Altogether the rat worked an hour and a half, pressing the lever 1,156 times before it, and the other two rats, had their hunger fully satisfied. One of the other rats worked the lever three times, but the third rat did not work it at all. Both soon became complete parasites on the "worker" rat.

A distinguished entomologist once said: "I can believe anything of an insect." Some of the results of the experiments with insects described in the previous pages may have inclined the reader to echo these words, and when he has read the following account of some insect endurance tests he will almost certainly agree with them.

The experiments were made by Dr. Frank E. Lutz, Curator of Entomology in the American Museum of Natural History. Lutz was interested in the heights to which insects and other small animals can be carried by winds above the earth. Glick, for example, by fixing collecting traps on an aeroplane, caught a spider at fifteen thousand feet, as well as numerous insects at lesser heights. Lutz wondered what was the maximum height to which insects, etc., could be carried and still survive. Could they, like man, survive a trip to the top of Mount Everest, where atmospheric pressure is less than one-third (225 mm.) that of sea-level (760 mm.)? It was not possible to duplicate entirely the conditions found in these frigid regions, but Lutz duplicated the two most important—decrease in atmospheric pressure and the consequent lack of oxygen.

A number of fruit-flies (*Drosophila*) were placed in a bell-jar and the air was gradually exhausted with a vacuum pump. A barometer inside the bell-jar indicated the pressure which had been reached. When this was equivalent to that at the top of Everest, the fruit-flies appeared to be behaving normally. More air was sucked out until the pressure inside the bell-jar was only 125 mm., the pressure of the air a third higher than Everest, or eight miles above sea-level. The flies were still walking about. No man, unless equipped with an oxygen mask and other mechanical aids, could live in an atmosphere so rarefied as that.

But still Lutz pumped air out of the bell-jar. He continued until the barometer was registering only 22 mm., which represented an altitude of seventeen miles above sea-level. The flies were not moving now but they were still alive and, when returned to normal conditions, they appeared none the worse for their drastic experience.

Lutz now carried the experiments a stage further. It is known that man can withstand moderate extremes in atmospheric pressure providing the changes are made slowly. But if they are made too quickly he suffers great pain, called "the bends," and if the change is very rapid, he dies. Lutz determined to see how the fruit-flies would reaction to sudden changes in atmospheric pressure.

Ten flies were placed in the bell-jar and the air was exhausted to the same point as in the previous experiments —22 mm. Then the valves in the apparatus were opened wide and pressure was almost instantaneously restored to normal—760 mm. Within four minutes all the flies were walking about as if such violent treatment were part of their daily routine.

Lutz repeated the same process a number of times. Not until after the eighth trial did one fly fail to recover within seven minutes. Two flies, a male and a female, survived twenty-four trials. Each trial, it should be remembered, consisted of taking the flies to a height of seventeen miles (speaking in terms of atmospheric pressure) and then almost instantaneously bringing them back to earth. No man could endure a quarter of one test, let alone twenty-four of them. Yet after the twenty-fourth trial Lutz put the two flies in a cage and they started to breed the next day. Careful microscopic examination of three generations bred from this pair failed to reveal anything unusual in their constitutions.

In another experiment Lutz put a cricket in a tank and pumped in air until the pressure was 2,300 mm., or nearly three times normal. For several minutes this pressure was maintained and then the stopper was removed and instantaneously pressure was restored to normal. The cricket paid no attention at all to these violent changes.

Lutz now determined to find out the ultimate in insect endurance. Three bees, two ants, a beetle and a grasshopper were placed in a glass tube. The ends of the tube were melted and welded into an apparatus which would transform the tube into a vacuum. The pump in this apparatus exhausted the air, until the pressure in the tube was of the order of one ten-thousandth of a millimeter.

Such a vacuum reproduced the conditions of inter-stellar space.

As the pump removed the air it also removed moisture, including that from the bodies of the insects. Towards the end of the experiment, the moisture which had been sucked out of the insects lay as a pile of snow on the floor of the apparatus. The insects thus risked death from desiccation as well as from the almost total vacuum in the tube.

The extreme vacuum was held for sixty seconds and then Lutz broke the glass tube with a hammer. The insects were thus instantaneously returned to normal atmospheric conditions, i.e. an increase in air pressure of over seven million times. Not an insect moved for some time, but two hours after the experiment they were all active and apparently in good condition. The next day one of the ants died, but all the other insects appeared to be quite normal.

Similar experiments were tried with a bumble-bee and two species of butterflies. They all survived the experiment, although on the following day the butterflies died. If a similar experiment were carried out with a man he would, of course, never survive the vacuum, but, if he did, then the instantaneous return to normal pressure would tear him to pieces.

Lutz comments on these experiments: "There was no longer room for doubt that insects are creatures that can not only completely recover within a few minutes from sudden and rapidly repeated transfers from the normal pressures to almost none and back again, but they are creatures that can survive the most complete vacuum that man can produce with exceptionally perfect apparatus. How do they do it; why can they do it?" It might be replied that the answers are to be found in the known facts about the anatomy and physiology of the insect respiratory and blood systems.

Lutz concludes his account of these experiments with his own characteristic answer to the question with which this chapter opened:

> What good are such experiments as these? Possibly collecting interesting information about the masterpieces of Creation is of no greater value than collecting human masterpieces of art; and possibly writing about Nature is no more useful than writing music; but, until someone is wise enough to be able to predict the worth of any bit of pure (as contrasted with "applied") science, we can at least say that it "amuses" those who do it and interests many who read about it.

12

How the Passenger Pigeon Became Extinct

"The greatest tragedy in Nature is the extinction of a species."
—ARTHUR A. ALLEN.

DURING its prime the passenger pigeon occurred in probably greater numbers than any other land bird the world has yet known. It ranged North America from the Atlantic seaboard to the Rocky Mountains, and from the Mackenzie district, in Canada, to the Gulf of Mexico. Incidentally, several specimens have been recorded for England, but they are generally believed to have escaped from aviaries.

A characteristic of the pigeon was its great massed flights, in which millions of the birds flew in tightly packed ranks. These flights were particularly noticeable during the migration seasons. There was a general northward migration in the spring and a southward one in the autumn. In addition, there were frequent flights to and from feeding, roosting and nesting sites.

There are many accounts from men who saw these flights and they agree closely enough to make us realise that there is no parallel to them in the world today. Here is an account of a migratory flight from the pen of William Wood, one of the early settlers in Massachusetts. Wood's book, published in 1635, is written with archaic spelling, but I have modernized it in this quotation:

> These birds come into the country to go to the north parts in the beginning of our spring, at which time (if I may be counted

worthy, to be believed in a thing that is not so strange as true) I have seen them fly as if the airy regiment had been pigeons, seeing neither beginning nor ending, length, or breadth of these millions of millions. The shouting of people, the rattling of guns, and pelting of small shot could not drive them out of their course, but so they continued for four or five hours together.

Pokagon, writing in 1895, but obviously referring to flights he had seen much earlier, says:

> I have seen them fly in unbroken lines from the horizon, one line succeeding another from morning until night, moving their unbroken columns like an army of trained soldiers pushing to the front, while detached bodies of these birds appeared in different parts of the heavens, pressing forward in haste like raw recruits preparing for battle.
>
> At other times I have seen them move in one unbroken column for hours across the sky, like some great river, ever varying in hue; and as the mighty stream sweeping on at sixty miles an hour, reached some deep valley, it would pour its living mass headlong down hundreds of feet, sounding as though a whirlwind was abroad in the land. I have stood by the grandest waterfall of America and regarded the descending torrents in wonder and astonishment, yet never have my astonishment, wonder and admiration been so stirred as when I have witnessed these birds drop from their course like meteors from heaven.

A more detailed account is the following, by King, writing of a flight he had seen about the middle of the nineteenth century, when, of course, the pigeons were already considerably reduced in numbers.

> Early in the morning I was apprised by my servant that an extraordinary flock of birds was passing over, such as he had never seen before. Hurrying out and ascending the grassy ramparts, I was perfectly amazed to behold the air filled, the sun obscured by millions of pigeons, not hovering about, but darting onwards in a straight line with arrow flight, in a vast mass a mile or more in breadth, and stretching before and behind as far as the eye could reach.

Swiftly and steadily the column passed over with a rushing sound, and for hours continued in undiminished myriads advancing over the American forests in the eastern horizon, as the myriads that had passed were lost in the western sky. It was late in the afternoon before any decrease in the mass was perceptible, but they became gradually less dense as the day drew to a close. At sunset the detached flocks bringing up the rear began to settle in the forest on the Lake-road, and in such numbers as to break down branches from the trees. The duration of this flight being about fourteen hours, the column (allowing a probable velocity of sixty miles an hour, as assumed by Wilson) could not have been less than three hundred miles in length, with an average breadth of one mile.

If the speed given was true ground speed, i.e. the pigeons were covering sixty land miles in an hour, the length of this flock would be 60 X 14 = 840 miles.

Audubon, writing of a flight he witnessed some forty years earlier, also refers to the sun being darkened by the pigeons' passing. He says:

The air was literally filled with pigeons; the light of noonday was obscured as by an eclipse, the dung fell in spots not unlike melting flakes of snow, and the continued buzz of wings had a tendency to lull my senses to repose . . . [at sunset] The pigeons were still passing in undiminished numbers and continued to do so for three days in succession.

This was not the effect they produced on other observers. Featherstonhaugh writes of flocks on the wing "creating whirlwinds as they move," and presenting "an image of the most fearful power. Our horse, Missouri, at such times, has been so cowed by them that he would stand still and tremble in his harness, whilst we ourselves were glad when their flight was directed from us."

In another passage Audubon said he found a "place of nightly rendezvous" which was about forty miles long with an average breadth of over three miles, its area being therefore over a hundred and twenty miles. Other writers speak of roosting areas even larger than this.

That other great biographer of American birds, Alexander Wilson, calculated the number in a flock which passed over him. He says the birds reached as far as the eye could see, both to right and left. He therefore put the breadth at a mile, although he believed, not unreasonably, that it was much more than this.

The flock took four hours to pass him and he assumed the birds were flying at sixty miles an hour. (There is independent evidence that the pigeons could fly at this speed.) The length of this flock, therefore, was about two hundred and forty miles. Wilson allowed three birds to each square yard, but as the flock was several strata deep, there were probably many more than this.

Wilson thus estimated that this flock contained 2,230,272,000 pigeons. Audubon saw another flight and estimated that 1,115,136,000 birds (exactly half Wilson's estimate) passed over and around him in three hours. In view of such numbers it is understandable that some of the early settlers in North America, after witnessing the passing of a flock, thought that it contained all the pigeons in America!

If such figures are accepted, and it seems reasonable to assume that they are correct in order of magnitude, then it means that in either of these two flocks there were several times more pigeons than there are land birds in the British Isles at the present time. It is obviously impossible to give anything but a very approximate figure for this number, but it does not appear to be greater than two hundred million. (Fisher.)

If half a pound is allowed for each bird, then the weight

of the flock Wilson saw was about half a million tons. Wilson computed that they would eat some seventeen million bushels of food a day. Such a quantity of food is far greater in bulk than would have been required for a day's ration for all the soldiers under arms in the world at the conclusion of the Second World War.

It may naturally be wondered what happened when one of these flocks alighted. Here is a description by Kalm, writing of a flock he saw in the spring of 1740.

> There came from the north an incredible multitude of these pigeons to Pennsylvania and New Jersey. Their number, while in flight, extended three or four English miles in length, and more than one such mile in breadth, and they flew so closely together that the sky and the sun were obscured by them, the daylight becoming sensibly diminished by their shadow. The big as well as the little trees in the woods, sometimes covering a distance of seven English miles, became so filled with them that hardly a twig or a branch could be seen which they did not cover; on the thicker branches they had piled themselves up on one another's backs, quite a yard high.
>
> When they alighted on the trees their weight was so heavy that not only big limbs and branches of the size of a man's thigh were broken straight off, but less firmly rooted trees broke down completely under the load. The ground below the trees where they had spent the night was entirely covered with their dung, which lay in great heaps. As soon as they had devoured the acorns and other seeds which served them as food and which generally lasted only for a day, they moved away to another place. . . .
>
> Several of the old men assured me that in the darkness they did not dare to walk beneath the trees where the pigeons were, because all through the night, owing to their numbers and corresponding weight, one thick and heavy branch after another broke asunder and fell down, and this could easily have injured a human being that had ventured below.

Kalm adds that several seafaring men said they had found areas at sea extending over three miles entirely covered with dead pigeons. It was believed that the pigeons had been carried away to sea during the storm, mist or snowfall, and then during the darkness of the night, possibly greatly fatigued, they had alighted on the water and been drowned.

What happened to the areas where the pigeons nested is vividly portrayed in the following passage by an obscure writer named Hinton, whose original work I have been unable to trace, but who is quoted by Buckingham.

> Their roosting-places are always in the woods, and sometimes occupy a large extent of forest. When they have frequented one of these places for some time, the ground is covered several inches deep with their dung; all the tender grass and underwood is destroyed; the surface is covered with large limbs of trees, broken down by the weight of the birds clustering one above another; and the trees themselves, for thousands of acres, killed as completely as if girdled with an axe.
>
> The marks of this desolation remain for many years on the spot; and numerous places can be pointed out, where, for several years afterwards, scarcely a single vegetable made its appearance.

Mease records that so many pigeons alighted on a hickory tree, over a foot in diameter, that its top was bent down to the ground and its roots started a little on the opposite side, so that a small ridge was raised in the soil. He adds that brittle trees were often broken down by the pigeons.

The pigeons nested three or four times a year and laid one or two eggs each time. As may be imagined, their nesting colonies were immense. They were generally in woods, and each tree normally contained many of the roughly-made nests. Over a hundred nests were sometimes built in a tree, and sometimes the branches broke under the weight of the rapidly-growing squabs.

In Petoskey, Michigan, in 1878, the nests occupied all except the smallest trees over an area twenty-eight miles long and between three and four miles wide. Wilson mentions another nesting colony in Kentucky which covered nearly twice the area of the Petoskey nests.

Before the settlers came to America the pigeons' chief foes were the elements and predatory animals, but neither of these agencies appears to have killed more birds than were replenished by the young each season.

The Indians also took their toll, but owing to their wise tradition of never killing during the nesting season, the great hosts remained undiminished. (Kalm.) It remained for the men from Europe to show the "ignorant savage" how to massacre the pigeons in their millions without any such nonsense as "close seasons."

The passenger pigeon played an important part in helping to feed some of the settlers in North America, particularly in Canada. Anburey, in his book of American travels published in 1789, says: "During the migratory flight of these pigeons, which generally lasts three weeks or a month, the lower sort of Canadians mostly subsist on them."

A Canadian settler, writing in 1832, says: "I think we should have half died if it had not been for the pigeons." (Quoted by Mitchell.) A further indication of the large part played by the pigeons in the diet of the Canadian settlers is found in the fact that in Ontario some of the servants stipulated that they would not eat pigeons more than so many times a week. Servants in Norway once made the same stipulation about salmon.

The birds were good eating and "pigeon pie" was a celebrated dish in former days. The feathers were used to stuff pillows and mattresses. The birds' gizzards were also used sometimes for medicinal purposes. According to an old recipe: "The gizzards dried and powdered were steeped and taken, an old-fashioned but reliable cure for vomiting stomach." (Mitchell.)

The pigeons' chief food was mast—acorns, beechnuts and chestnuts. These the immense virgin forests of North America provided in abundance. But the pigeons were not averse to other food and when, in the spring, they came across one of the settlers' freshly-sown fields, or in the autumn upon the ripe grain, they made short work of the farmer's labors.

Great numbers of birds came to these feasts and sometimes the shocks of grain were so covered with pigeons that nothing else could be seen. A man might sow his field in the morning, go home to lunch, and find on his return that the pigeons had also had their lunch—on his fields, and not a seed was left of those which had been so laboriously sown a few hours before. The only compensation for the victims of these depredations were a few more pigeon pies.

The pigeons were great gluttons and they sometimes ate until their crops literally burst. Margaret H. Mitchell, of the Royal Ontario Museum of Zoology, in her exhaustive work on the passenger pigeon, quotes a correspondent as saying some pigeons "cleaned up" his father's pea field and, with their crops filled with dry peas they "flew to the stream near their nesting-place and drank water. The dry peas swelled and either burst their crops or choked them."

Townsend says from the first the settlers began to destroy the pigeons excessively. In view of the need to obtain an additional meat supply, and also to protect their crops, such destructiveness is understandable. It would doubtless have sounded fantastic if anyone had suggested to the colonists that a time would come, and that in the not too distant future, when of the countless hosts of pigeons that darkened their skies with their flights, not one would be left alive.

Although they undoubtedly accounted for great numbers of pigeons, it was not the settlers who caused the wholesale slaughter which eventually led to the extermination of the species. This responsibility rests with the "pigeoners," men who had discovered that money could be made out of the pigeons, and whose sole business was killing and marketing them.

During the boom period, in the 1860's, it was estimated that there were five thousand men in the country whose sole business was pigeons. They earned anything from ten to fifty dollars a day during the breeding season, which normally lasted from March to July.

This commercial killing of the pigeons began in the 1840's, and continued, with increasing destructiveness, to its peak, about 1870. This "perfection of extermination," as Townsend calls it, was aided by the opening up of the country. The cutting down of forests tended to make the pigeons concentrate and thus rendered it easier for their butchers to find them, the telegraph system enabled the pigeoners to be advised of the location of flocks to be slaughtered, and the railways enabled the dead birds to be carried quickly to market.

To anyone with the slightest concern for wild-life, the accounts of the orgies of pigeon slaughter make sickening reading. Every method which the ingenuity of man could devise for killing, or capturing the pigeons alive, was ruthlessly exploited.

The birds were shot by every weapon capable of discharging a missile into their serried ranks. The Indians were experts with bow and arrow, but white men also occasionally used them to save powder and shot. Guns, rifles, blunderbusses, pistols and revolvers of every type, date and calibre were fired into the dense flocks as they flew overhead, or when they alighted on the ground or in trees.

Sometimes swivel-guns, a type of cannon charged with a pound or more of small shot and a good charge of powder, were used, and one discharge of this weapon sometimes killed sufficient birds to feed a small community.

When a flock once flew near a fort on Lake Ontario, the soldiers loaded a cannon with grape shot and discharged it at the pigeons. As a result hundreds of birds fell into the lake and were picked up by men who put off from the shore in boats.

Dunlop, in his book published in 1832, recounts the following diverting episode of what happened when a large flock of the pigeons flew over Toronto.

> For three or four days the town resounded with one continuous roll of firing, as if a skirmish were going on in the streets—every gun, pistol, musket, blunderbuss and fire-arm of whatever description, was put in requisition. The constable and police magistrate were on the alert, and offenders without number were *pulled up*—among whom were honourable members of the executive and legislative councils, crown lawyers, respectable, staid citizens, and last of all, the sheriff of the county; till at last it was found that pigeons, flying within easy shot, were a temptation too strong for human virtue to withstand; and so the contest was given up, and a sporting jubilee proclaimed to all and sundry.

The following account by Welford (quoted by Lowe) gives a good account of one man's experience of pigeon shooting during the spring flight of 1870:

> The birds came over in incredible numbers, some idea of which may be gained by what happened to me personally. I was up that

morning very early, and so were the birds. I had taken up a position on the top of some rising ground, behind a rail or small fence which ran along the edge of a wood in which were growing some beech trees, which supplied the favourite food of the pigeons. The beech-nuts had been lying covered with snow all through the winter, but were now exposed. Between the spot where I stood and another large wood was a small open clearing or meadow. By this time the air was black with flock upon flock of pigeons all going eastward. Some were flying high, but others just cleared the wood in front of me, and then, swooping down to the meadow, flew very close to the ground, so close indeed that it was necessary for them to rise before clearing the low fence in front of me. This was my opportunity; and as they cleared the fence, so I fired into wave upon wave.

They came on in such numbers that thousands would pass between the discharge of my double-barrelled gun and its reloading —a longer process then, in the days of muzzle-loading, than now. [1922.] At about 10 a.m., not being in the least prepared for such phenomenal slaughter, I ran out of powder and shot, having then four hundred birds to my credit, during the shooting of which it was not unusual to get from fifteen to twenty-five with a "right and left." Being now unable to do any more shooting until I had secured more ammunition, I hurried home, a distance of one and a half miles, got a horse and light wagon, returned to the scene of my battue with some grain-bags holding one and a half bushels of ordinary grain, filled them with the pigeons and made tracks for my home again. All the time I was filling the sacks the birds were still streaming low over the fence, so that before leaving I hid myself behind it, and taking a long slender cedar rail, knocked down many more as they came over.

This, however, to my then youthful notions, did not appeal so much as shooting, so that, after dropping my birds at home, I drove into town (Woodstock, Ontario) for more powder and shot and caps, a distance of three and a half miles. During the entire drive there and back, millions of pigeons were filling the air and shadowing the sun like clouds. The roar of their wings resembled low rumbling thunder, and the shooting from scores of guns could be heard for miles, resounding from wood to wood, like a small mimic battle. This great flight continued from before daylight to dusk, and lasted for some days, gradually lessening until the flight was over.

An interesting point about passenger pigeon shooting was that a shot at birds flying towards the marksmen often failed because the shot glanced off the thickly-feathered, highly-polished breasts. But a shot as the birds were flying away would bring them tumbling down. (Fleming.)

In what may be called legitimate shooting, as opposed to the commercial battues, there was real sport. Margaret H. Mitchell says: "From the sportsman's viewpoint wild pigeons were fine game. A bright October day with a hint of frost in the air, a good old muzzle-loader in hand, and pigeons flying, and what more could man's heart desire? The bird's swiftness on the wing and its rolling, twisting flight made it no easy target, and considerable skill was needed to bring down an individual."

Many were the tales told of large bags at single or double discharges—of hand guns, that is, I am not now dealing with bags made by 18-pounders and similar engines of destruction. It was not unusual to bag a dozen at a shot. In fact, one old-timer said where he lived guns were seldom fired unless it was fairly sure five or six pigeons would be killed by the one shot. Sometimes a man would manoeuvre until he had a number of pigeons in a fairly straight line, on a fence, for example, and then, by shooting them at an angle, he could strip off the whole line. A dozen or more birds thus succumbed to the one shot.

Another method used to ensure a heavy bag with a single shot was to dig a long trench and scatter grain in it to attract the pigeons. On finding this banquet the birds crowded into the trench and filled it to capacity. A hidden marksman would then discharge into the trench an old flint-lock musket loaded with a handful of shot. Seventy-five birds were thus sometimes secured with the one shot.

One man claimed to have bagged ninety-nine pigeons

with one shot. When asked why he did not make it a hundred, he said he would not tell a lie for one small pigeon! (Mitchell.) A correspondent of Pennant's said: "Sir William Johnson told me that he once killed at one shot with a blunderbuss, a hundred and twenty or thirty."

The highest figure I have seen was reputed to have been made in 1662, when a man, evidently wielding a fearful blunderbuss, is said to have accounted for a hundred and thirty-two pigeons with one shot. (Thwaites.)

Whatever may be thought of the authenticity of these records, it certainly seems to be true that occasionally over a hundred pigeons were brought down by a right and left. Etta S. Wilson says her father, who was a famous pigeon shot, used a muzzle-loading, double-barrelled shot gun, and he often brought down seventy birds with the double discharge. "But I have heard my mother say that she once saw him bring down a hundred and twenty-four birds at one shot, and she was a truthful woman." Mershon refers to a man who killed a hundred and forty-four pigeons with two barrels of a six-bore shoulder gun. And Kalm says several people told him a man at Schenectady had killed a hundred and fifty pigeons with two discharges of bird-shot.

In addition to shooting, many other methods were employed to destroy the pigeons. Sometimes flocks flew so low that pigeons were knocked out of the sky by men armed with long poles, and when tired birds were flying very low over water men in boats sometimes knocked them down with oars. The denseness of the flocks made such things possible.

"Pigeons had a habit of sometimes flying very low over hill tops, barn roofs or other eminences, and people often posted themselves at such points of vantage and knocked the birds down with clubs, poles, or branches." (Mitchell.) But at other times the pigeons flew so high that only a bullet from a good rifle could reach them.

Boys were paid to climb trees in which flocks were roosting at night and knock the pigeons down with sticks. One boy said he knocked down three hundred pigeons in a night, and that figure does not appear to have been unusual. Sleeping birds were also knocked out of trees by men on the ground using long poles. Iron pots containing burning sulphur, and torches of pine knots helped to stupefy the birds and add to the slaughter.

Pigeons were trapped and snared in various ways. Sometimes a boy would surround himself with sheaves of grain, thus forming a sort of tent, and when pigeons alighted on them he would catch birds by the legs and pull them inside. Box traps were used, some of them as large as a wagon. Mather says one man surprised "two hundred dozen" in his barn, where the pigeons were eating his grain. He shut the door and caught the lot!

So dense were the ranks of the pigeons in some of their flights that almost any sort of missile thrown at them brought down at least one pigeon. A boy was carrying a pair of sheep shears when a flock began to pass over his head. He threw the shears into the air and down came a pigeon. Brickbats, clubs and stones thrown into a low-flying flock frequently brought down one or more pigeons.

Etta S. Wilson writes: "We had a big watch-dog, of reputable mixed bull and shepherd descent, an efficient custodian of our premises, but no hunter. When pigeon-time came he was always the first to reach the hill, whining with excitement and jumping crazily around in a frenzy of expectation. I have seen him spring up into the air as a low-flying flock came upward and grab a bird on the wing."

Great numbers were taken by netting, a method perfected by the pigeoners who often wanted live birds to send away for trap-shooting. A piece of ground would be baited and a net, some six feet wide and thirty feet long, would be fixed in position. "Stool pigeons," birds tied to a pole that could suddenly be raised and lowered by a cord, or "fliers," captive birds thrown into the air, were used to lure down flocks of pigeons flying overhead. When sufficient birds were under the net, it would be sprung, and a haul of dozens or even hundreds of pigeons would be captured. A double net once captured 1,332 pigeons with a single throw, and as many as 5,000 were caught by one net in a day. One man captured 24,000 pigeons in ten days. (Roney.)

The worst feature of the whole pigeon trade was the massacre of the squabs. As already mentioned, the pigeons used to nest in great colonies, and in the following passage Audubon describes how one of these, said to be forty miles long, was attacked in Kentucky:

> As soon as the young were fully grown, and before they left the nests, numerous parties of the inhabitants, from all parts of the adjacent country, came with wagons, axes, beds, cooking utensils, many of them accompanied by the greater part of their families, and encamped for several days at this immense nursery. . . .
>
> The ground was strewed with broken limbs of trees, eggs, and squab pigeons, which had been precipitated from above, and on which herds of hogs were fattening. Hawks, buzzards, and eagles were sailing about in great numbers and seizing the squabs from their nests at pleasure; while, from twenty feet upwards to the tops of the trees, the view through the woods presented a perpetual tumult of crowding and fluttering multitudes of pigeons, their wings roaring like thunder; mingled with the frequent crash of falling timber, for now the axe-men were at work cutting down those trees that seemed to be most crowded with nests, and contrived to fell them in such a manner that in their descent they might bring down several others; by which means the falling of

one large tree sometimes produced two hundred squabs, little inferior in size to the old ones, and almost one mass of fat. . . .

It was dangerous to walk under the flying and fluttering millions, from the frequent fall of large branches, broken down by the weight of multitudes above, and which in their descent often destroyed numbers of the birds themselves; while the clothes of those engaged in traversing the woods were completely covered with the excrements of the pigeons.

The numbers of pigeons killed or taken alive by all these methods were, of course, enormous. Roney visited the great Petoskey nesting colony in 1878 and writes as follows of what he saw, and of the number of birds which were sent to market.

> Scarcely a tree could be seen but contained from five to fifty nests, according to its size and branches. Directed by the noise of chopping and of falling trees, we followed on, and soon came upon the scene of action. Here was a large force of Indians and boys at work, slashing down the timber and seizing the young birds as they fluttered from the nest. As soon as caught, the heads were jerked off from the tender bodies with the hand, and the dead birds tossed into heaps. Others knocked the young fledglings out of the nests with long poles. . . .
>
> For many weeks the railroad shipments averaged fifty barrels of dead birds per day—thirty to forty dozen old birds and about fifty dozen squabs being packed in a barrel. Allowing five hundred birds to a barrel, and averaging the entire shipments for the season at twenty-five barrels per day, we find the railroad shipments to have been 12,500 dead birds daily, or 1,500,000 for the summer. Of live birds there were shipped 1,116 crates, six dozen per crate, or 80,352 birds. These were railroad shipments only and not including the cargoes by steamer from Petoskey, Cheboygan, Cross Village, and other like ports, which were as many more. Added to this were the daily express shipments in bags and boxes, the wagon loads hauled away by the shot-gun brigade, and the myriads of squabs dead on the nests.

Martin, in an article written to refute Roney's figures as exaggerated, would concede only one and a half million as the total destruction of pigeons from all sources!

When it is remembered that these figures refer to only one nesting colony, it will be seen that the total number of pigeons being killed annually in this country and Canada at this time (1878) must be reckoned in tens, if not hundreds, of millions. No species, however prolific, could long continue to suffer at that rate and not be in danger of extinction. Some legislative efforts were made to prevent that calamity but, as has happened with other threatened species in the past, really effective legislation came too late. Seeing the apparently inexhaustible numbers of the pigeons, people laughed at the idea of laws to protect them. In fact in 1848 a law was passed in Massachusetts imposing a fine on anyone interfering with the netting of pigeons.

Three years later the first law favourable to the pigeons was passed in Vermont. This law of 1851 protected all non-game birds all the year round. Other laws dealing directly with the passenger pigeon were later passed in various states, but they were usually ignored and seldom enforced.

In 1880 flocks of pigeons of considerable size were still occasionally reported, although they were, of course, not comparable to the vast hosts seen a few decades earlier. But in 1900, only twenty years later, the species was, for all practical purposes, extinct.

The causes which led to the extinction of this, the most prolific species of land bird in the world, make an interesting study, as well as providing a valuable object lesson for all conservationists. Margaret H. Mitchell points out that although it died out with relative suddenness, its extinction was not so rapid as some of the old-time pigeoners believed—for example, that it "went out like dynamite."

The reason for such rapid extinction appears to be this.

For each animal there seems to be a minimum number below which the species cannot continue to exist. No one appears to have given a satisfactory explanation of why such a result should follow serious numerical reduction, but it is almost as if the species as a whole lost its will to live.

On this point Townsend says:

> A bird accustomed for ages to living together in large numbers and close ranks, whether in feeding, migrating, roosting, or nesting, might find it impossible to continue satisfactorily these functions with greatly reduced and scattered ranks. It is probably no mere figure of speech to say that under these circumstances such a communistic bird would "lose heart," nor is it fanciful to suppose that sterility might in consequence affect the remnants. [See also Allee.]

Although there may be some mystery about the final disappearance of the passenger pigeon, there is no difficulty in finding reasons why its numbers fell below the danger-line. But first it is of interest to look at some of the fanciful reasons advanced for its extermination.

One theory was that the birds were all drowned, various waters being suggested as their grave, from the Gulf of Mexico to the Atlantic Ocean. Thousands of the birds *were* drowned occasionally in bad weather while crossing the Great Lakes, but this never occurred in sufficient numbers appreciably to affect the species as a whole. Some people said the birds, presumably to escape the incessant persecution, migrated to South America, or even to Australia. Another wild theory was that they flew to the North Pole and were there frozen to death.

To anyone who reads the history of the passenger pigeon the causes which led to its decline are all too obvious. Many millions of the birds were killed each year by man and Nature, and the constant harrying of the nesting areas and massacre of the squabs prevented adequate replacements. Any species can become extinct if a sufficiently large proportion of its offspring is killed each year before they reach maturity. The everyday hazards of Nature cut off the rest. Snow and ice occurring during some of the nesting seasons towards the end of the century probably hastened the now inevitable end.

A few birds lived on in captivity for some years after the last wild passenger pigeon was dead. The last of these captive birds, and the last member of its species, died in the Cincinnati Zoological Gardens, on September 1, 1914.

From time to time there are reports that passenger pigeons have been seen alive. Generally, these are mourning doves which, at a distance, are very similar to the passenger pigeon. While most responsible ornithologists consider the species became extinct in 1914, a note in *The Passenger Pigeon,* for April, 1944, indicates that there *may* be one or two stragglers still alive.

13

How the Buffalo Was Saved

THE AMERICAN bison was the most numerous game animal civilized man has known. Not even the teeming herds of game that flourished in South Africa before the coming of the Boer could equal the spectacle presented by the bison when first seen by Western eyes. They lived and travelled as have no other four-footed beasts, moving in great multitudes, like grand armies in review, and covering scores of square miles at a time.

Bison ranged over a third of the continent, from Northern Mexico to the northern shore of the Great Slave Lake in Canada. Their favourite resort was the Great Plains of the Middle West. Here they were seen in their millions. Here, too, grew the "buffalo grass" they loved so well and, owing to constant grazing, whole tracts of country in this area appeared to be covered with a smooth green carpet.

(The American buffalo is, of course, not a buffalo but a bison. Strictly speaking the word buffalo should be restricted to the Asiatic water buffalo and the African buffalo.)

Pedro de Castañeda, who travelled across the Great Plains during the middle of the sixteenth century, said there were so many bison that he did not know what to compare them with for multitude except the fish in the sea. Other early explorers used such phrases as "countless herds," "incredible numbers" and "teeming myriads" in endeavours to convey the impression made on their minds by these vast hosts of animals.

Seton has made a careful estimate of how many bison existed in North America before the great herds began to be attacked by the white man, and considers the figure must be between fifty and sixty millions. Garretson says that a conservative estimate would "place the number easily at sixty millions."

Such figures convey little, but accounts of those who saw the bison in its prime help us to realize something of the sight they presented. Capt. Bonneville, writing of a journey he made in 1832 to the Rocky Mountains, said that in the north fork of the Platte River he ascended a high bluff commanding an extensive view of the surrounding plains. He writes: "As far as my eye could reach, the country seemed absolutely blackened by innumerable herds." (Quoted by Irving.)

Similarly Townsend, writing of a journey in the same region at about the same time as Bonneville, says:

> Towards evening, on the rise of a hill, we were suddenly greeted by a sight which seemed to astonish even the oldest among us. The whole plain, as far as the eye could discern, was covered by one enormous mass of buffalo. Our vision, at the very least computation, would certainly extend ten miles, and in the whole of this great space, including about eight miles in width from the bluffs to the river bank, there was apparently no vista in the incalculable multitude.

Garretson says he once asked an old plainsman and bison hunter how many of these animals he had seen at one time. The man said:

> Picture in your mind an open grassy valley a mile wide and straight for many miles, level as a floor, bare of any trees or brush, and on each side bluffs stretching away east and west in parallel lines to the horizon. Early one morning in 1851 I stood on an eminence overlooking this valley; and from bluff to bluff on the

north and on the south, and up the valley to the westward—as far as the eye could reach—the broad valley was literally blackened by a compact mass of buffalo, and not only this—the massive bluffs on both sides were covered by thousands and thousands that were still pouring down into the already crowded valley, and as far as the eye could reach, the living dark masses covered the ground completely as a carpet covers the floor. It looked as if not another buffalo could have found room to squeeze in, and a man might have walked across the valley on their huddled backs as on a floor.

Seton (1927) points out that it was only during the migration seasons that the very large herds were seen.

Colonel R. I. Dodge wrote that in May, 1871, he drove a light wagon between two forts on the Arkansas River. For at least twenty-five miles he travelled through one immense herd which was migrating to the north. He says:

The whole country appeared one great mass of buffalo, moving slowly to the northward; and it was only when actually among them that it could be ascertained that the apparently solid mass was an agglomeration of innumerable small herds, of from fifty to two hundred animals, separated from the surrounding herds by greater or less space, but still separated. The herds in the valley sullenly got out of my way, and, turning, stared stupidly at me, sometimes at only a few yards' distance. When I had reached a point where the hills were no longer more than a mile from the road, the buffalo on the hills, seeing an unusual object in their rear, stared an instant, then started at full speed directly towards me, stampeding and bringing with them the numberless herds through which they passed, and pouring down upon me all the herds, no longer separated, but one immense compact mass of plunging animals, mad with fright, and as irresistible as an avalanche.

The situation was by no means pleasant. Reining up my horse (which was fortunately a quiet old beast that had been in at the death of many a buffalo, so that their wildest, maddest rush only caused him to cock his ears in wonder at their unnecessary excitement), I waited until the front of the mass was within fifty yards, when a few well-directed shots from my rifle split the herd, and sent it pouring off in two streams to my right and left. When all had passed me they stopped, apparently perfectly satisfied, though thousands were yet within reach of my rifle and many within less than a hundred yards. Disdaining to fire again, I sent my servant to cut out the tongues of the fallen. This occurred so frequently within the next ten miles, that when I arrived at Fort Larned I had twenty-six tongues in my wagon, representing the greatest number of buffalo that my conscience can reproach me for having murdered on any single day.

From this and other accounts it is known that this host of bison took about five days to pass a given point and was some fifty miles deep. As Dodge says it was at least twenty-five miles wide it is estimated that it contained over four million animals!

The trails made by the bison during their periodical travels have left indelible marks across the face of the North American continent. Garretson says the bison was the best natural engineer the world has ever known, because it invariably followed the path of least resistance and chose the easiest grade possible in the direction and through the country it was travelling. The early surveyors, who sought a way for the railroad across the plains, followed the old bison trails for mile after mile without being able to improve the grade.

Hulbert, in his book on America's historic highways, comments on this aspect of the bison as follows:

It is very wonderful that the buffalo's instinct should have found the very best courses across a continent upon whose thousand rivers such great black forests were thickly strung. Yet it did, and the tripod of the white man has proven it, and human intercourse will move constantly on paths first marked by the buffalo. It is interesting that he found the strategic passageways through the mountains; it is also interesting that the buffalo marked out the most practical paths between the heads of our rivers, paths that are closely followed to-day by great railroads.

Before the colonization of America the worst foe of the bison was the elements. Blizzards, cutting down visibility to a few feet and covering the ground with several feet of snow, killed many. Some herds appear to have been entirely destroyed by blizzards, and thousands of bison from one herd were killed when they fell over a precipice during one of these violent snow-storms. (Garretson.) Snow that thawed and then froze into a hard crust was another hazard, as it prevented the bison from grazing, and if the thaw was delayed, as it sometimes was, for many days thousands of bison died of cold and starvation.

Victor H. Cahalane comments on this paragraph: "From my observations of bison in South Dakota, I question that they pay much attention to blizzards as such. They can forage easily in snow up to twenty-four inches deep. Thirty inches causes difficulty, but is certainly not fatal. Of course, a crust that endures for weeks is another matter. A stampeding herd might plunge over a cliff in a blizzard, but ordinarily the bison is a phlegmatic and canny animal about such matters."

Treacherous spring ice on the rivers was responsible for drowning large numbers. Seton, in fact, says this was the greatest of all destroyers in the days before man became their arch-foe. A herd would commence to cross a wide ice-bound river. The ice would support the bison for a time, but as more and more of the great creatures (they averaged about half a ton) swarmed across, the ice would give and bison would start crashing through and be drowned. So great would be the press of bison at the rear that those who had seen the fate of their fellows in front would be powerless to turn back, and they, in turn, would be drowned. As many as 7,360 drowned and mired bison were counted in a day in and along the banks of one river. (Henry.)

But not all the danger came in the winter and spring. Summer brought drought and sometimes it was so severe that all the smaller streams dried up. The bison then travelled many miles seeking water and many died on the way. With summer came also the dreaded prairie fires, and there can have been few more pitiable sights than thousands of bison swept by a sea of fire. Sometimes the long shaggy coats would catch fire and the tortured beasts would plunge like living torches madly through the herd. Often the fire would blind many of them, and after it had passed they would run heedlessly about, mad with pain and a thirst they were unable to quench.

The Great Plains are one of the worst areas for tornadoes, the most terrible of all wind-storms. It is impossible to imagine the scene when the long, dark funnel of violently whirling air swept across the closely packed bison herds. Bison would be tossed to right and left like chaff, and a swathe of dead and dying animals be cut which would be several miles long and perhaps half a mile wide. So great is the power of the updraught in a tornado that some bison would be lifted from the ground and carried through the air—possibly for miles if the tornado were very violent. Two bison were once discovered completely stripped of hair and with every bone in their bodies crushed. They had evidently been carried aloft and then dashed to earth by a tornado. (Rollins, Moore, and Lane.)

There was only one animal which habitually preyed on the bison, and that was the wolf. They followed the herds in great packs, destroying many calves, and the old and feeble beasts on the outskirts. Mead said he believed each wolf on the plains killed a dozen bison a year, including calves. The bison fought back and many a wolf met its death when it approached too near to those powerful

horns. Sometimes a young calf would be entirely surrounded by a number of adults, which faced a circle of
wolves intent on capturing the calf. It took a very hungry
and desperate wolf to endeavour to break through such a
protective cordon.

Sometimes the wolves would surround a bison and keep
on attacking it until it succumbed. Frequently the wolves
would start to eat it before it was dead. Inman has the following description of one such attack.

> We saw standing below us in the valley an old buffalo bull,
> the very picture of despair. Surrounding him were seven grey
> wolves in the act of challenging him to mortal combat. The poor
> beast, undoubtedly realizing the hopelessness of his situation,
> had determined to die game. His great shaggy head, filled with
> burrs, was lowered to the ground as he confronted his would-be
> executioners, his tongue, black and parched, lolled out of his
> mouth, and he gave utterance at intervals to a suppressed roar.
>
> The wolves were sitting on their haunches in a semicircle im-
> mediately in front of the tortured beast, and every time that the
> fear-stricken buffalo gave vent to his hoarsely modulated groan,
> the wolves howled in concert in most mournful cadence.
>
> After contemplating his antagonists for a few moments, the
> bull made a dash at the nearest wolf, tumbling him, howling,
> over the silent prairie; but while this diversion was going on in
> front, the remainder of the pack started for his hind legs to ham-
> string him. Upon this the poor beast turned to the point of attack,
> only to receive a repetition of it in the same vulnerable place by
> the wolves, who had as quickly turned also and fastened them-
> selves on his heels again. His hind quarters now streamed with
> blood, and he began to show signs of great physical weakness. He
> did not dare to lie down; that would have been instantly fatal.
> By this time he had killed three of the wolves, or so maimed them
> that they were entirely out of the fight.
>
> At this juncture the suffering animal was mercifully shot, and
> the wolves allowed to batten on his thin and tough carcase.

Bison were the Indians' wild cattle. The flesh was very
similar to beef and could be eaten for a longer period than

that of any other animal without becoming monotonous. But, in addition to being his main food supply, the bison provided the Indian with numerous by-products. Garretson, in his excellent survey of the history of the bison, summarizes these uses in the following passage.

> The hide was dressed in a variety of ways, each special treatment having its particular use. The robe was the Indian's winter covering and his bed, while the skin, freed from the hair and dressed, constituted his summer blanket. The dressed hide was used for moccasins, leggings, shirts and women's dresses. Dressed buffalo cow skins formed his lodge. Braided strands of rawhide served him as ropes, which were also made from twisted hair. The green hide was often used as a pot in which to boil meat, also in making the round bull boats for crossing rivers. The tough, thick hide of the bull's neck was employed in war shields that would turn an arrow or a musket ball. From the rawhide, with the hair shaved off, were constructed parfleches—a sort of folding box or envelope for holding dried meat and small articles.
>
> The ribs made runners for small sleds drawn by dogs. The hooves were used for making glue for fastening the feathers and heads on arrows, and also for rattles; the hair for cushions and saddle pads; the long, black hair and beard for ornaments on wearing apparel, shields and quivers. Water buckets were made from the lining of the paunch. Part of the flat shoulder-blade, tied to a beaded string, constituted the "bull-roarer" which, when whirled rapidly about, reproduced the grunt accompanied by the tom-toms and flutes in the buffalo dance. From the sinews along the back, thread and bow-strings were made. The horns were made into spoons and the ornaments of war bonnets, and even bows.

The Indians used several methods to kill bison. Sometimes a small semi-circular space would be fenced off with tree trunks below a perpendicular bank, which was shallow enough for the bison to jump down but too steep for them to scale. Bison would then be driven over the bank and find themselves trapped. The Indians then fired at and speared the imprisoned animals, many of which were killed by the horns and hooves of their fellows in their frantic struggles to escape.

Another method sometimes used was to lure the animals over a river precipice, some of which extended for miles. An Indian would fasten a bison skin about his body and wear a bison's horns and ears on his head. He then placed himself between a herd and a precipice. Other Indians would then frighten the herd and start it running towards the disguised Indian, who started running towards the precipice. Seeing him in the distance, the herd followed, for they were accustomed to following a leader, and their limited intelligence prevented them from distinguishing between a real bison and the skin, horns and ears of a bison running on a two-legged human body.

As soon as the Indian who was acting as the decoy reached the river precipice, he hid himself in a crevice of the cliff which he had selected beforehand. The leading bison thus led to the brink of the precipice were forced over by the rush of the closely-packed herd thundering along behind, which could, of course, see no danger in front but only that from the pursuing Indians in the rear. Hundreds of bison might thus be drowned or crushed by the fall before the rest of the herd realised the danger. By such a ruse the Indians, with little trouble, gained a large store of meat and other commodities which the bison supplied.

The Indians occasionally set fire to the grass around a herd and advanced to the slaughter behind the wall of flames. Herds were also attacked by Indians on horseback, who rode at the animals from all directions and caused them to mill about in wild confusion, goring each other and trampling one another to death, the Indians meanwhile adding to the wild mêlée by spearing and shooting at the frantic beasts. In winter, when snow lay thick on the ground, the Indians went hunting on snowshoes. The heavy bison, floundering helplessly about in the deep snow, were then easy prey for the Indians' arrows and lances. Bison swimming across rivers were sometimes speared and shot.

The most sporting method was called "running buffalo," in which animals were chased on horseback and the kill made at close quarters. Prior to the coming of the Spanish explorers, the Indians knew nothing of horses, but once they started to use them they soon became the finest horsemen in the world. Racing at full speed, they could jump from one horse to another and keep the sport up hour after hour. Holding their bodies on one side of a horse, they would discharge a flight of arrows from under its neck.

A horse intended for the bison hunt was carefully trained and was never used for any other purpose. So fully did these animals enter into the spirit of the chase that if an Indian fell off during a hunt his horse would sometimes continue the chase on its own. One horse, being left in the camp after the others had left on a hunt, broke out and took part in the hunt as best it could with no rider on its back.

Hornaday thus describes how the hunt was conducted:

> Whenever the hunters discovered a herd of buffalo, they usually got to leeward of it and quietly rode forward in a body, or stretched out in a regular skirmish line, behind the shelter of a knoll, perhaps, until they had approached the herd as closely as could be done without alarming it. Usually the unsuspecting animals, with a confidence due more to their great numbers than anything else, would allow a party of horsemen to approach within from two hundred to four hundred yards of their flankers, and then they would start off on a slow trot. The hunters then put spurs to their horses and dashed forward to overtake the herd as quickly as possible. Once up with it, each hunter chooses the best animal within his reach, chases him until his flying steed carries him close alongside, and then the arrow or the bullet is

sent into his vitals. The fatal spot is from twelve to eighteen inches in circumference, and lies immediately back of the fore-leg, with its lowest point on a line with the elbow.

This, the true chase of the buffalo, was not only exciting, but dangerous. It often happened that the hunter found himself surrounded by the flying herd, and in a cloud of dust, so that neither man nor horse could see the ground before him. Under such circumstances fatal accidents to both men and horses were numerous. It was not an uncommon thing for half-breeds to shoot each other in the excitement of the chase; and, while now and then a wounded bull suddenly turned upon his pursuer and overthrew him, the greatest number of casualties were from falls.

The following account by Ross (1856), writing of a great hunt in which about four hundred hunters took part, gives some idea of the hazards involved in "running buffalo": "On this occasion the surface was rocky and full of badger holes. Twenty-three horses and riders were at one moment all sprawling on the ground; one horse, gored by a bull, was killed on the spot; two more were disabled by the fall; one rider broke his shoulder-blade; another burst his gun and lost three of his fingers by the accident; and a third was struck on the knee by an exhausted ball. These accidents will not be thought over-numerous, considering the result, for in the evening no less than thirteen hundred and seventy-five tongues were brought into camp."

The horse of a famous Comanche hunter ran for four hours at high speed among a great herd of bison. Its meandering course was estimated to cover nearly fifty miles, during which the hunter lanced a hundred bison. A Kiowa chief, who was over seven feet tall, always chased bison on foot and lanced them to death.

To be charged by an enraged bison must have been a terrifying experience, for their strength is prodigious. A cow once charged a horseman and lifted horse and rider on to its horns. It then carried them a hundred yards at full speed until it collided with a fence, over which they were tossed. On another occasion a bull was being driven up a chute into a railway wagon. The bull charged straight ahead and went through the side of the wagon as though it was made of paper. (Ellsworth.)

Despite the losses inflicted by both natural causes and the Indians, until the coming of the white settlers the bison not only held its own, but appears actually to have increased in numbers. This was the position until about the middle of the seventeenth century, when the white man began to invade their eastern haunts. Civilization and the bison could not live side by side. Gradually, as the settlers advanced farther and farther west, ever greater numbers of the bones of slaughtered bison were left to whiten the lands where once the great herds roamed.

As the westward emigration continued through the eighteenth and into the nineteenth century, more and more bison were killed. The settlers now realized that it was a commercially valuable animal and, in addition to killing many themselves, they also encouraged the Indians to do so by offering them fire-water (whisky, often poisonous) and gunpowder for the hides and tongues of bison. The Indians, thus equipped with firearms and with the urge to slaughter more, were themselves partly responsible for killing off the hosts on which their ancestors lived for centuries.

So vast, however, were the herds on the Great Plains and some little-colonized regions, that even as late as the middle of the nineteenth century there were still millions of bison left. But then, in the 1860's, there began the building of the Union Pacific transcontinental railway. It ran eventually from Chicago to San Francisco, right through the heart of the bison country. Henceforward, the bison to the north of the railway were known as the northern herd and those to the south as the southern herd.

The bison did not take kindly to the invasion of their centuries-old territory by the noisy steel monsters. If a train stood between a herd and its destination, so much the worse for the train. The bison charged and threw the train off the tracks. Often the couplings between carriages were broken and the engines were crippled.

Often trains were held up while enormous herds crossed the tracks. At first the engine-drivers tried to charge through, only to have their engines derailed and damaged. After a few such experiences the bison were given the right of way. One train was held up for two hours while a herd three miles wide crossed the line. In such circumstances the passengers usually amused themselves by shooting indiscriminately into the massed ranks of the bison. After this herd had passed, for example, not fewer than five hundred carcasses lay on either side of the train.

The railways provided a ready means of transport for the hides, meat and other products of the slaughtered bison. The professional butchers and others who earned their living from the bison flocked to the kill, and as the steel lines advanced farther and farther west across the greatest refuge of the bison, there began the most enormous and wasteful slaughter of game animals the world has ever known.

Tens of thousands alone were killed to help feed the men constructing the railway. The railway officials employed professional hunters. One man killed 1,500 bison in seven days, his record for a day being 250. Garretson says the best authenticated total for a season's kill was 3,300, the season in this instance running from September 1 of one year to April 1 of the next. William F. Cody was one of the professional hunters and thus earned his world-famous title of "Buffalo Bill." In eighteen months he killed 4,280 bison. Once he shot 69 in eight hours.

It is sometimes said that "Buffalo Bill" is a misnomer, because Cody never killed a buffalo in his life; only bison. But this statement overlooks the fact that he did once shoot a true buffalo, an Indian water buffalo, in a circus, that was injured and had to be put out of its misery. (Garretson.)

The world has never known more wasteful hunting than occurred with the bison. Many a lone hunter killed a bison for a single meal, and frequently only the tongues were taken from slaughtered animals. Catlin says a party of several hundred Sioux Indians slew 1,400 bison and took only their tongues, which they traded for a few gallons of traders' whisky. When the railway ran across the Great Plains excursions were advertised for trippers to shoot the bison from the trains just for "sport."

Bison were hunted on horseback in a similar way to that practised by the Indians. But the white hunters used either large-bore Colt revolvers, sometimes one in each hand, or a repeating rifle. "Kit" Carson, a famous guide and trapper, once killed seven bison with six bullets by riding into a herd with his special "buffalo gun." After he had shot his sixth bison with his sixth shot he noticed a projection under the skin of the dead animal. On examination this proved to be Carson's bullet, so he cut it out, reloaded his rifle with powder and cap and, inserting this bullet, killed another bison! (Garretson.)

The most popular and deadly method employed by the professional killers was known as the still hunt. Equipped with a Sharp's or a Spencer's rifle, weighing upwards of twenty pounds and firing a half-inch calibre bullet, the hunter would creep on all fours to within two or three hundred yards of the leeward side of a herd, if possible taking refuge in a slight depression or behind rocks or other obstacles.

From this position the hunter would shoot the most suspicious sentinel. If the herd ran he ran with them, but generally they soon stopped and the same process would begin again. Eventually, by continuously shooting any animal which assumed the leadership, the herd could be made to stand, and then the hunter picked off animal after animal, the rest merely looking stupidly on as they saw their fellows falling to the ground.

Writing of this method of killing bison, Hornaday says:

> The policy of the hunter is to not fire too rapidly, but to attend closely to business, and every time a buffalo attempts to make off, shoot it down. One shot per minute was a moderate rate of firing, but under pressure of circumstances two per minute could be discharged with deliberate precision. With the most accurate hunting rifle ever made, a "dead rest," and a large mark practically motionless, it was no wonder that nearly every shot meant a dead buffalo. The vital spot on a buffalo which stands with its side to the hunter is about a foot in diameter, and on a full-grown bull is considerably more. Under such conditions as the above, which was called getting "a stand," the hunter nurses his victims just as an angler plays a big fish with light tackle and in the most methodical manner murders them one by one, either until the last one falls, his cartridges are all expended, or the stupid brutes come to their senses and run away. Occasionally the poor fellow was troubled by having his rifle get too hot to use, but if a snow-bank was at hand he would thrust the weapon into it without ceremony to cool it off. . . .
>
> One hunter of my acquaintance, Vic Smith, the most famous hunter in Montana, killed one hundred and seven buffaloes in one "stand," in about one hour's time, and without shifting his point of attack. . . . Where buffaloes were at all plentiful, every man who called himself a hunter was expected to kill between one and two thousand during the hunting season—from November to February—and when the buffaloes were to be found it was a comparatively easy thing to do.

Sometimes the killers—they do not observe the honoured name of hunter—would form a line of camps along the watering places of the bison and also set guards over the water-holes. As the bison came to slake their thirst during the burning summer hours they would be greeted with a fusillade of shots. At night fires were lit so that the men could see to shoot down any thirst-maddened beast that tried to break through the armed cordon under cover of darkness. A whole herd could be wiped out in this way, and frequently was.

So can man sink below the level of the beast he slaughters.

The years of greatest destruction appear to have been 1872-74. Hornaday calculated that the number of bison slain in the southern herd during this period was 3,698,730. Dodge, presumably including the bison slain in both southern and northern herds, gives a figure for the same period of 5,373,730. By the end of the hunting season of 1875 the great southern herd, which had comprised the greatest collection of game animals in the world, had practically ceased to exist.

Most of the few thousand survivors of the countless millions which had once thundered over the grassy plains of the Middle West, fled to the south-west and dispersed through the wild, inhospitable country about the Texas Panhandle and the Pecos River. Here, and in a few other wild spots, the bison managed to live for several years, but men who had known no other trade save bison hunting, sought them out in their last strongholds, and by 1889 there were probably less than seventy bison left south of the Union Pacific railroad.

The extermination of the northern herds was not long delayed. As the Northern Pacific railway extended its lines northward the hunters followed, and the orgy of slaughter which had already exterminated the southern herds was repeated. By 1889 there were probably less than twenty wild bison left in the United States north of the Union Pacific.

In addition to the eighty or ninety wild bison living in the United States at this date there were 456 living in captivity or under Government protection, and some 550 of the so-called wood bison living in Canada. Thus, of the forty or so million bison living in the whole of North America at the turn of the century, there were little over a thousand left eighty-nine years later. The actual figure given by Hornaday of all bison alive on January 1, 1889, is 1,091. Of this number, 85 in the United States and 550 in Canada were running wild and unprotected. Seton (1927) considers the minimum number was reached in 1895, when probably not more than 800 bison were left in the whole of North America. And only at this late stage was public opinion sufficiently aroused for determined action to be taken to prevent the disappearance forever of the greatest and most valuable game animal on the American continent.

Several attempts had been made earlier in the century to pass legislation to protect the fast-dwindling herds, but the measures had been shelved and no decisive action was taken. It was left to private individuals to take the first practical steps to save the American bison from being exterminated. In 1873 a Pend d'Orielle Indian, named Walking Coyote, captured two male and two female calves. He protected and cared for them and they began to breed. In 1884 the little herd comprised thirteen head and in that year they were sold to two ranchers named Charles Allard and Michel Pablo. In 1896 the herd numbered some three hundred head.

Several other public-spirited men who realised that if the bison became extinct something typically American, and one of the most vivid links with pioneer days, would vanish forever from the American scene, captured wild bison calves and protected them on their ranches. There was also a small herd of wild bison in the Yellowstone National Park, in Wyoming. The Canadian herd of wood bison in Athabaska was now under Government protection and was supervised by the North-west Mounted Police.

This was the position in May, 1894. A law was then rushed through Congress which made it illegal for anyone to kill bison or other game animals in the Yellowstone National Park. (Poachers had already reduced this, the last herd of wild bison in the United States, to about twenty head.) Hereafter anyone killing a bison in the Park was liable to a fine of one thousand dollars or imprisonment.

Victor H. Cahalane comments on this paragraph: "The two States (actually territories) of Idaho and Wyoming prohibited hunting prior to 1894. Of course, the Federal Act of March 1, 1872, which established Yellowstone National Park, afforded protection to the bison and other animals, 'against wanton destruction.' The Act of 1894 merely put teeth in the prohibition."

With this new national realisation of the precarious state of the bison the five hundred-odd left alive in the country in 1889 just managed to hold their own until the turn of the century. Then, in 1902, a number of bison from the herds kept by private ranchers were purchased and placed in the Yellowstone National Park. This herd, with a few calves from the wild bison already there, was placed in special corrals and was protected as no other bison, and probably no other large animals, have ever been before or since.

Writing of this herd, Cahalane (October, 1944) says:

Every individual animal was carefully watched, guarded and practically prayed over. Shelters were provided against bad weather. All dangerous natural instincts were suppressed. Cows were segregated from bulls except for breeding. Calves were removed from their mothers at an early age to assure against accidental injury. Some were bottle-nursed. Every precaution was taken to prevent the loss of a single animal.

Later other bison reserves were established. In 1905 the American Bison Society was founded and received the blessing of President Theodore Roosevelt in its work to preserve and increase the bison, and protect North American big game generally. Slowly the bison came back from the brink of annihilation. In 1908 there were 2,047 in the whole of North America, or nearly double the number living at the time of Hornaday's census in 1889. A census was taken nearly every subsequent year and each one showed an increase. In 1933 the figure was 21,496: the bison was saved.

To-day the country has more bison than it can cope with. The Government herds have periodically to be thinned out. Some animals are given away. Anyone who has adequate facilities can obtain a bison free, providing he pays the freightage and undertakes to fulfil a few simple conditions. (Smith.) Other surplus bison are shot and the meat is given to the Indians. Under domestication the bison have begun to lose some of the vigorous characteristics of their ancestors. The artificial feeding and other controls of the Yellowstone herd are, therefore, being dropped in an endeavor to help them regain their native virility.

It is hoped eventually to establish a herd on the Great Plains. Thus the descendants of the men who once brought the bison to the verge of extinction may see Old Man Buffalo once again roaming at liberty over his true home range.

BIBLIOGRAPHY

Bibliography

1. The Split Second in Nature

Amery, L. S. *Days of Fresh Air* (1939).

Bent, Arthur Cleveland. *Life Histories of North American Diving Birds* (1919); *Life Histories of North American Wild Fowl* (1923 and 1925); *Life Histories of North American Birds of Prey* (1937 and 1938). (U.S. National Museum.)

Burpee, Royal H., and Stroll, Wellington. "Measuring Reaction Time of Athletes," *Research Quarterly of the American Physical Education Association*, March, 1936.

Burr, Malcolm. *The Insect Legion* (1939).

Buytendijk, F. J. J. *Reaktionszeit und Schlagfertigkeit* (1932).

Buytendijk, F. J. J., Fischel, W., and ter Lagg, P. B. "Über den Zeitsinn der Tiere," *Arch. Néerl. Physiol.* (1935), vol. 20, p. 123.

Caldwell, O. H. "Testing Munitions by Radio Magic," *Science Digest*, August, 1941.

Daly, Marcus. *Big Game Hunting and Adventure, 1897–1936* (1937).

Dawson, William L. *The Birds of California* (1923).

Ditmars, Raymond L. *The Fight to Live* (1938).

Elbel, E. R. "A Study of Response Time Before and After Strenuous Exercise," *Research Quarterly of the American Physical Education Association*, May, 1940.

Evans, C. Lovatt. *Starling's Principles of Human Physiology* (1941— eighth edition, edited and revised by Evans).

Forbush, E. H. *A History of the Game Birds, Wild-Fowl and Shore Birds of Massachusetts and Adjacent States* (1912).

Free, E. E. "What Can Happen While You Wink," *Popular Science Monthly*, August, 1928.

Gill, E. L. "Crows, Rooks and Starlings *versus* Kestrels and Peregrine Falcons," *British Birds*, June, 1919.

Gladstone, Hugh S. *Record Bags and Shooting Records* (1930).

Gnanamuthu, C. P. "Comparative Study of the Hyoid and Tongue of Some Typical Genera of Reptiles," *Proceedings of Zoological Society*, Ser. B, April, 1937.

Gosse, Philip H. *The Birds of Jamaica* (1847).

277

HARDY, MANLY. "The Otter," Forest and Stream, March 11, 1911.

JENNISON, MARSHALL W. "The Dynamics of Sneezing—Studies by High-Speed Photography," Scientific Monthly, January, 1941.

KORTRIGHT, FRANCIS H. The Ducks, Geese and Swans of North America (1943).

LESTER, E. C. "Are You Accident Prone?" The Autocar, March 1, 1940.

MAGNAN, A. Le Vol des Insectes (1934).

MURPHY, ROBERT CUSHMAN. "The Most Amazing Tongue in Nature," Natural History, May, 1940.

NORTH, ESCOTT. The Saga of the Cowboy (1942).

O'CONNOR, JACK. "Splitting the Seconds," Esquire, September, 1945.

PARR, G. "Human Relay System," Wireless World, March 10, 1938.

PERRY, RICHARD. At the Turn of the Tide (1938).

PITT, FRANCES. Woodland Creatures (1922).

POHL, VICTOR. Bushveld Adventures (1940).

POPE, CLIFFORD H. Amphibians and Reptiles of the Chicago Area (1944).

PRIEST, CECIL D. The Birds of Southern Rhodesia (1933).

ROBERTSON, I. A. Letter in The Field, April 6, 1940.

ROSENBLITH, W. A., and COX, RICHARD T. Quoted in Science Digest, April, 1941.

ROSTAND, JEAN. Toads and Toad Life (1934).

ROTH, CHARLES B. "The Myth of the Two-Gun Man," American Mercury, October, 1937.

ST. JOHN, CHARLES. Wild Sports and Natural History of the Highlands (1927).

SAVAGE, WILLIAM. "Cooper's Hawk," Western Ornithologist (1900), vol. 5, p. 6.

SEASHORE, SIGFRID H., and SEASHORE, ROBERT H. "Individual Differences in Simple Auditory Reaction Times of Hands, Feet and Jaws," Journal of Experimental Psychology, October, 1941.

SETON, ERNEST THOMPSON. Lives of Game Animals (1927).

SMITH, MALCOLM A. "Cobras and King Cobras," Zoo Life, Summer, 1946.

SUTHERLAND, ROBERT. Zambesi Camp Fires (1935).

TEALE, EDWIN WAY. The Lost Woods (1946).

THOMAS, CHAUNCEY. Quoted in Field and Stream, August, 1936.

VERRILL, A. HYATT. Strange Reptiles and Their Stories (1937).

WILLIAMSON, H. D. "In the Sun" (lizards), Walkabout, March, 1941.

ZOOND, A. "The Mechanism of Projection of the Chameleon's Tongue," Journal of Experimental Biology, April, 1933.

2. Fish With an Ear for Music

Abbott, Chas. C. "The Intelligence of Fish," *Science*, April 27, 1883.

Allee, W. C. *The Social Life of Animals* (1938).

Berridge, W. S. *All About Fish* (1933).

Boulenger, E. G. *Queer Fish* (1925).

Bower-Shore, Clifford. "Character in Fish," *John o' London's Weekly*, December 25, 1936; *Character in Fish* (1938).

Brown, Frank A. "Responses of the Large-Mouth Black Bass to Colors," *Bulletin Illinois N.H. Survey* (1937), vol. 21, p. 33.

Bull, H. O. "Studies on Conditioned Responses in Fishes," *Journal of Marine Biological Association* (1928), vol. 21, p. 33.

Clarke, W. J. Quoted in *Country Life*, February 1, 1941.

Donne, T. E. *Rod Fishing in New Zealand Waters* (1927).

Farrington, S. Kip. *Pacific Game Fishing* (1942).

French, John W. Quoted in *Science Digest*, November, 1941.

Frolov, J. P. *Fish Who Answer the Telephone* (1937); *Pavlov and His School* (1937).

Grey, Zane. *Tales of Fishing Virgin Seas* (1925).

Gudger, E. W. "Some Instances of Supposed Sympathy Among Fishes," *Scientific Monthly*, March, 1929.

Herter, Konrad. "Dressurversuche an Fischer," *Zeitschr. vergl. Physiologie* (1929), vol. 10, p. 688.

Hewitt, Edward R. *Secrets of the Salmon* (1925).

Hill, Craven. *Evening Standard*, June 5, 1934.

Holder, C. F., and Jordon, D. S. *Fish Stories* (1909).

Keay, A. E. Letter in *John o' London's Weekly*, October 14, 1938.

Martin, Edward A. Letter in *John o' London's Weekly*, October 28, 1938.

Moore, William. "Odd Fish," *The Listener*, February 22, 1940.

Parker, G. H. "Hearing and Allied Senses in Fish," *Bulletin U.S. Fish Commission* (1902), vol. 22, p. 45; "The Sense of Hearing in Fishes," *American Naturalist* (1903), vol. 37, p. 185.

Pincher, H. Chapman. "Hearing in Fishes," *Angling*, April–June, 1940; "Are Fish Intelligent?", *Country Life*, November 9, 1945.

Russell, E. S., and Bull, H. O. "A Selected Bibliography of Fish Behaviour," *J. du Conseil int. Explor. Mer* (1932), vol. 7 (2), p. 255.

Spooner, G. M. "The Learning of Detours by Wrasse," *Journal of Marine Biological Association*, March, 1937.

Stetter, H. "Untersuchungen über den Gehörsinn der Fische, besonders von *Phoxinus laevis* L. und *Ameiurus nebulosus* Raf.", *Zeitschr. vergl. Physiologie* (1929), vol. 9, p. 339.

VASQUEZ, RAUL. Quoted in *Field and Stream*, February, 1937.

VON FRISCH, KARL. "The Sense of Hearing in Fish," *Nature*, January 1, 1938.

WARNER, LUCIEN H. "The Problem of Color Vision in Fishes," *Quarterly Review of Biology*, September, 1931.

WASHBURN, MARGARET F. *The Animal Mind* (1936).

WHITE, GERTRUDE M. "Association and Color Discrimination in Mudminnows and Sticklebacks," *Journal of Experimental Zoology*, February, 1919.

WOLFF, H. "Das Farbensunterscheidungsvermögen der Elritze," *Zeitschr. vergl. Physiologie* (1925), vol. 3, p. 279.

3. DANGER! FLYING BIRDS

ANONYMOUS. "Radar for Airports and Planes," *Science*, April 6, 1945.

AXILROD, B. M., and KLINE, G. M. "Resistance of Transparent Plastics to Impact," U.S. National Advisory Committee for Aeronautics Technical Note No. 718 (1939).

BEADNELL, C. M. "The Toll of Animal Life Exacted by Modern Civilization," *Proceedings of Zoological Society*, Series A, July, 1937.

BRAMLEY, ERIC. "Bird Menace Reported Serious in India-Burma Theatre," *American Aviation*, June 1, 1945.

BROOKS, MAURICE. "Electronics as a Possible Aid in the Study of Bird Flight and Migration," *Science*, March 30, 1945.

BROWN, V. H. "Aircraft Collision with a Goldfinch," *Auk*, January, 1945.

BUSS, IRVEN O. "Bird Detection by Radar," *Auk*, July, 1946.

CARR-LEWTY, R. A. "Reactions of Birds to Aircraft," *British Birds*, January, 1943.

CHAPIN, JAMES P. "Wideawake Fair Invaded," *Natural History*, September, 1946.

CURTIN, PAT. "Duck, Brother, Duck!", *Air Transport*, October, 1944.

Daily Express, October 19, 1942.

DAY, J. WENTWORTH. "Birds and the Bomber," *The Listener*, March 13, 1941; "Airmen's Encounters with Birds," *Country Life*, May 24, 1941.

DE LABILLIERE, F. C. "Peril to Pilots," *Field*, December 1, 1945.

DUDLEY, JOHN. "Birds *versus* Aeroplanes," *Daily Mirror*, April 5, 1935.

EISERER, LEONARD. "Bird-proof Windscreens Expected Soon," *American Aviation*, December 1, 1941.

GLADSTONE, HUGH S. *Birds and the War* (1918).

HAMEL, GUSTAV, and TURNER, CHARLES C. *Flying* (1914).

LACK, DAVID. *The Life of the Robin* (1943).

LACK, DAVID, and VARLEY, G. C. "Direction of Birds by Radar," *Nature*, October 13, 1945.

LANE, FRANK W. "Birds *versus* Aircraft," *Discovery*, 1950.

Life, October 13, 1941, August 10, 1942, November 23, 1942, March 1, 1943.

MAKIN, WILLIAM J. *African Parade* (1934).

MARKHAM, BERYL. *West with the Night* (1943).

MORSE, A. L. *Science News Letter*, March 21, 1942; "New Windshield Developments," *S.A.E. Journal* (*Transactions*), August, 1943.

News-Chronicle, September 21, 1935; August 14, 1941; October 19, 1942, December 8 and 9, 1942; August 1, 1947.

PIGMAN, GEORGE L. *Impact-Resistant Windshield Construction* (1943). (Published by the U.S. Civil Aeronautics Administration.)

PITMAN, C. R. S. *A Game Warden Takes Stock* (1942).

SEVENTY-SEVENTH CONGRESS (Second Session) OF THE HOUSE OF REPRESENTATIVES OF THE U.S., Report No. 2383 (1942).

SUPF, PETER. *Airman's World* (1933).

TERRES, JOHN K. "Birds Have Accidents Too!", *Audubon Magazine*, January–February, 1946.

WALPOLE-BOND, JOHN. *A History of Sussex Birds* (1938).

WILKINS, H. T. *Mysteries of the Great War* (1935).

WILLIAMS, JOHN R. "Kingbird Attacks Airplane," *Auk*, January, 1935.

4. ANIMAL ACCIDENTS

ALLEN, GLOVER MORRILL. *Bats* (1939).

AMERY, L. S. *Days of Fresh Air* (1939).

AUSTIN, OLIVER L., and AUSTIN, OLIVER L., junior. "Food Poisoning in Shore Birds," *Auk*, April, 1931.

BEADNELL, C. M. "The Toll of Animal Life Exacted by Modern Civilization," *Proceedings of Zoological Society*, Series A., July, 1937.

BENT, ARTHUR CLEVELAND. *Life Histories of North American Gallinaceous Birds* (1932); *Life Histories of North American Birds of Prey* (1937 and 1938); *Life Histories of North American Cuckoos, Goatsuckers, Hummingbirds and Their Allies* (1940); *Life Histo-*

ries of North American Flycatchers, Larks, Swallows and Their Allies (1942). (U.S. National Museum.)

BERG, CARLOS. "Una Araña Pescadora," *Anales Sociedad Cientifica Argentina* (1883), vol. 15, p. 245.

BOULENGER, G. A. *The Tailless Batrachians of Europe* (1897).

BRIDGES, T. C. *Wardens of the Wild* (1937).

BROWNE, C. A. R. "A Bird Killed by a Mantis," *Journal of Bombay Natural History Society*, July, 1899.

BRYDEN, H. A. *Enchantments of the Field* (1930).

BURR, MALCOLM. *The Insect Legion* (1939).

BURROUGHS, JULIAN. "A Chimney Swift Invasion," *Bird-Lore* (1922), vol. 24, p. 210.

CHALMERS, PATRICK. *The Anglers' England* (1938).

CHOLMONDELEY-PENNELL, H. *The Angler-Naturalist* (1863).

CUMING, E. D. *Idlings in Arcadia* (1936).

CUNNINGHAM, D. D. *Plagues and Pleasures of Life in Bengal* (1907).

DAWSON, WILLIAM L. *The Birds of California* (1923).

DEWAR, DOUGLAS. "How Animals Die," *Indian Field*, December 22, 1904; *Indian Bird Life* (1925).

DREYER, W. A. "The Question of Wildlife Destruction by the Automobile," *Science*, November 8, 1935.

EALAND, C. A. *Animal Ingenuity of To-day* (1921).

EVERMANN, B. A. *A List of the Birds Obtained in Ventura County, California* (1886), vol. 3, p. 179.

FISHER, ARTHUR H. Quoted in *Science Digest*, October, 1941.

FISKE, WILSON B. "Wildlife at the Crossroads," *Nature Magazine*, May, 1940.

FORBUSH, EDWARD HOWE. *Birds of Massachusetts and Other New England States* (1927).

FOWKE, PHILIP. "Jungle Tragedies," *Chambers's Journal*, May, 1936.

GIBBINGS, ROBERT. *Sweet Thames Run Softly* (1940).

GLADSTONE, HUGH S. *Record Bags and Shooting Records* (1930).

GLEGG, WILLIAM E. "Fishes and other Aquatic Animals Preying on Birds," *Ibis*, July, 1945.

GOIN, OLIVE B. "Peromyscus Impaled on Opuntia," *Journal of Mammalogy*, May, 1944.

GUDGER, E. W. "Foreign Bodies Found Embedded in the Tissues of Fishes," *Natural History*, September–December, 1922; "More About Spider Webs and Spider Web Fish Nets," *Bulletin New York Zoological Society*, July, 1924; "Spiders as Fishermen and Hunters," *Natural History*, May–June, 1925; "Wide-Gab, the Angler Fish," *Natural History*, March–April, 1929; "Rubber Bands on

Mackerel and Other Fishes," *Annals and Magazine of Natural History,* November, 1929; "Voracity in Fishes," *Natural History,* November–December, 1929; "Some More Spider Fishermen," *Natural History,* January–February, 1931; "More Spider Hunters," *Scientific Monthly,* May, 1931; "Coelenterates as Enemies of Fishes," *Annals and Magazine of Natural History,* February, 1934; "Fishes and Rings," *Scientific Monthly,* December, 1937; "The Fish in the Iron Mask," *Scientific Monthly,* March, 1938; "Some Ctenophore Fish Catchers," *Scientific Monthly,* July, 1943; "Fishermen Bats of the Caribbean Region," *Journal of Mammalogy,* February, 1945.

GURNEY, J. H. *The Gannet* (1913).

HALL, ERNEST. "A Woodpecker Mystery," *The Field,* August 24, 1935.

HAMILTON, W. J., junior. "Small Mammals Trapped by Plants," *Journal of Mammalogy,* February, 1939.

HARLOW, W. M. "Trapped," *Nature Magazine,* February, 1935.

HARTING, JAMES E. *Recreations of a Naturalist* (1906).

HAUGEN, ARNOLD O. "Highway Mortality of Wildlife in Southern Michigan," *Journal of Mammalogy,* May, 1944.

HOPKINS, H. C. "Oyster *versus* Kingfisher," *Ornithologist and Oologist* (1892), vol. 17, p. 109.

HUBBARD, WYNANT DAVIS. "How African Animals Die," *Chambers's Journal,* April, 1937.

HUDSON, W. H. *The Book of a Naturalist* (1919).

JOHNSON, PAUL B. "Accidents to Bats," *Journal of Mammalogy,* May, 1933.

JOKL, ERNST. *Aviation Medicine* (1942).

JOY, NORMAN H. "Fatal Collision of Swifts," *British Birds,* November, 1930.

KEARTON, RICHARD. *At Home with Wild Nature* (1922).

KIPLING, JOHN LOCKWOOD. *Beast and Man in India* (1892).

KOLB, C. HAVEN, junior. "Meadow Mouse Trapped by Plant Stalks," *Journal of Mammalogy,* August, 1939.

KRUMM-HELLER, ARNOLDO. "Die Feinde der Klapperschlange," *Kosmos* (1910), vol. 7, p. 417.

LEY, WILLY. "The Pit that Swallowed Monsters," *Animal and Zoo Magazine,* March, 1939.

LINCOLN, FREDERICK C. "Some Causes of Mortality Among Birds," *Auk,* October, 1931; *The Migration of North American Birds* (1935), Circular No. 363 of the U.S. Department of Agriculture.

LOCKWOOD, MARY E. "Hummingbird and Bass," *Bird-Lore* (1922), vol. 24, p. 94.

LOW, JESSOP B. "Coots Killed Under Unusual Circumstances," *Wilson Bulletin,* June, 1934.

LOWE, WILLOUGHBY P. "Bird and Snail Associations," *Ibis,* October, 1943; January, 1944.

LUTTRINGER, LEO A., junior. "Dramas of the Wild," *American Forests,* November, 1938.

MACINTYRE, DUGALD. "How Birds Combine Speed with Safety," *Chambers's Journal,* November, 1935.

MAKIN, WILLIAM J. *African Parade* (1934).

McKEOWN, KEITH C. *Spider Wonders of Australia* (1936).

McNEILL, FRANK. "Birds of a Tropic Isle," *Australian Museum Magazine,* July–September, 1946.

MILLER, ALDEN H. "Tribulations of Thorn-dwellers," *Condor,* September, 1936.

MILLS, ENOS A. *In Beaver World* (1913).

MITCHELL, MARGARET H. "The Passenger Pigeon in Ontario" (1935), *Contribution No. 7 of the Royal Ontario Museum of Zoology.*

MORGAN, BANNER BILL. "Bird Mortality," *Passenger Pigeon,* April, 1944.

NICHOLS, CHARLES K. "A Peculiar Injury to a Robin," *Auk,* July, 1944.

NICHOLS, DAVID G. "Varied Thrush Trapped by Acorn," *Condor,* May–June, 1940.

OSMASTON, B. B. "Owl Caught on a Thorn," *Journal of Bombay Natural History Society,* October, 1916.

PANGBURN, CLIFFORD. "Extraordinary Fatality to a Blue-winged Teal," *Auk,* January, 1945.

PARKER, ERIC. *Oddities of Natural History* (1943).

PETERS, T. M. "A Spider Fisherman," *American Naturalist* (1876), vol. 10, p. 688.

PETTINGILL, OLIN SEWALL, junior. "King Rail Impaled on Barbed Wire," *Auk,* October, 1946.

PITMAN, C. R. S. *A Game Warden Takes Stock* (1942).

PITT, FRANCES. "Starlings That Were Tied Together," *Evening News,* December 14, 1942.

POHL, VICTOR. *Bushveld Adventures* (1940).

PRIOR, SOPHIA. *Carnivorous Plants and 'The Man-Eating Tree'* (1939), Leaflet No. 23 of the Field Museum of Natural History, Chicago.

PROTHEROE, ERNEST. *New Illustrated Natural History of the World* (1936).

RAND, A. L. "A Modern Bird Fatality," *Auk*, July, 1938.

ROGERS, C. GILBERT. "A Natural Bird-lime," *Journal of Bombay Natural History Society*, October, 1911.

ROSS, ROLAND CASE. "Do Black Phœbes Eat Honey-bees?" *Condor* (1933), vol. 35, p. 232.

SETON, ERNEST THOMPSON. *Lives of Game Animals* (1927).

SHADLE, ALBERT R. "A Crow Impaled in Flight," *Auk*, April, 1931; "The Attrition and Extrusive Growth of the Four Major Incisor Teeth of Domestic Rabbits," *Journal of Mammalogy*, February, 1936.

SHAW, WILLIAM T., and CULBERTSON, A. E. "A Flock of Cedar Waxwings Meets Tragedy," *Condor*, July, 1944.

SHELLEY, LEWIS O. "Tree Swallow Tragedies," *Bird-Banding* (1934), vol. 5, p. 134.

STAGER, KENNETH E. "California Leaf-nosed Bat Trapped by Desert Shrub," *Journal of Mammalogy*, August, 1943.

STÄHLIN, ADOLF. "Feststellung der Todesursache von Haustieren und Wild," *Arb. Thür. Landesanstalt f. Pflanzenbau u. Pflanzenschutz*, Jena, part 3 (1944).

STELFOX, A. W. "Diptera and Other Insects Speared by Grass," *Entomologist's Monthly Magazine*, March, 1946.

STONER, DAYTON. "Wildlife Casualties on the Highways," *Wilson Bulletin*, December, 1936.

SUTTON, GEORGE MIKSCH. "An Unfortunate Pine Warbler," *Auk*, April, 1938.

TERRES, JOHN K. "Birds Have Accidents Too!", *Audubon Magazine*, January–February, 1946.

THOMSON, J. ARTHUR. *The New Natural History* (1926).

TWIST, R. F. "Trapped by a Tree," *The Field*, September 16, 1944.

VENABLES, L. S. V. Letter in *Journal of Mammalogy*, August, 1944.

WALPOLE-BOND, JOHN. *A History of Sussex Birds* (1938).

WARD, J. D. U. "Birds Killed by Wires," *Discovery*, October, 1937.

WILLETS, G. V. "Reducing Bird Electrocution," *Nature Magazine*, January, 1946.

WILLIAMS, C. B. "An Insect-catching Grass," *Entomologist*, March, 1945.

WITHERBY, H. F. "Curious Fatality to a Redbreast and its Young," *British Birds*, August, 1932.

YATES, C. A. *Kruger National Park* (1935).

5. STRANGE USES FOR ANIMALS

AKNAY, MARGOT. Quoted in *Parade*, February, 1938.

ALEXANDER, W. B. *Birds of the Ocean* (1928).

ALLEN, GLOVER MORRILL. *Bats* (1939).

BANFIELD, E. J. *Tropic Days* (1918).

BATES, C. A. Quoted in *American Magazine,* March, 1940.

BEEBE, WILLIAM. *The Edge of the Jungle* (1935).

BURR, MALCOLM. *The Insect Legion* (1939).

CORNER, E. J. H. "Wayside Trees of Malaya," *Hong Kong Naturalist* (1941), vol. 10, Nos. 3 and 4.

CUMING, E. D. *Idlings in Arcadia* (1936).

DEANE, WILLIAM. *Fijian Society* (1921).

FOX, H. MUNRO. Selene (1928).

GUDGER, E. W. "The Puffer Fishes and Some Interesting Uses of Their Skins," *Bulletin New York Zoological Society,* November, 1919; "Monkeys Trained as Harvesters," *Natural History,* May–June, 1923; "More About Spider Webs and Spider Web Fish Nets," *Bulletin New York Zoological Society,* July, 1924; "A New Purgative, the Oil of the 'Castor Oil Fish,' Ruvettus," *Boston Medical and Surgical Journal,* January 15, 1925; "Stitching Wounds With the Mandibles of Ants and Beetles," *Journal of American Medical Association,* June 13, 1925; "Helmets from Skins of the Porcupine-Fish," *Scientific Monthly,* May, 1930; "Strange Stories of Fish," *Scientific Monthly,* December, 1942; "The Giant Fresh-Water Fishes of South America," *Scientific Monthly,* December, 1943.

GUPPY, H. B. *The Solomon Islands and Their Natives* (1887).

HOLDER, CHARLES F. *Living Lights* (1887).

HOPKINSON, E. "The Frigate-Bird Post of the Pacific," *Bulletin New York Zoological Society,* July, 1919.

KIPLING, JOHN LOCKWOOD. *Beast and Man in India* (1892).

KNIGHT, F. C. E. "Cockfighting in Bali," *Discovery,* February, 1940.

MANN, WILLIAM M. "Monkey Folk," *National Geographic Magazine,* May, 1938.

McKEOWN, KEITH C. *Insect Wonders of Australia* (1935); *Spider Wonders of Australia* (1936).

MEEK, A. S. *A Naturalist in Cannibal Land* (1913).

METCALF, C. L., and FLINT, W. P. *Destructive and Useful Insects* (1939).

MOODIE, ROY L. "The Chub and the Texas Horn Fly," *American Naturalist,* March, 1909.

MURPHY, ROBERT CUSHMAN. *Oceanic Birds of South America* (1936).

NORRIS, THADDEUS. "The Use of Bats," *American Sportsman* (1874), vol. 4, p. 90.

PRATT, A. E. *Two Years Among New Guinea Cannibals* (1906).

RATCLIFFE, FRANCIS. *Flying Fox and Drifting Sand* (1938).

SANDERSON, IVAN T. *Living Treasure* (1941).

STERNDALE, R. A. *Mammalia of India* (revised edition by FRANK FINN, 1929).

STEVENSON, CHARLES H. "Utilization of the Skins of Aquatic Animals," *Annual Report of the U.S. Commission of Fish and Fisheries for* 1902 (1904).

SWAN, JAMES G. "The Eulachon or Candle-fish of the Northwest Coast," *Smithsonian Miscellaneous Collection for* 1880 (1881).

TEALE, EDWIN WAY. *The Golden Throng* (1942).

TURNER, GEORGE. *Samoa a Hundred Years Ago and Long Before* (1884).

VERE, PERCY. "Hedgehogs, and a New Use for Their Spines," *Chambers's Journal*, June, 1917.

VERRILL, A. HYATT. *Strange Insects and Their Stories* (1937).

WILKINSON, SIR GARDNER. *Manners and Customs of the Ancient Egyptians* (1879).

WILLIAMSON, ROBERT W. *The Ways of the South Sea Savage* (1914).

YOUNG, CAPT. WILLIAM E. *Shark! Shark!* (1933).

6. WHY ANIMALS ACT "DUMB"

BATTEN, H. MORTIMER. *Inland Birds* (1923).

BEEBE, W. *Tropical Wild Life in British Guiana* (1917).

BONNIER, G. "Le sens de la Direction chez les Abeilles," *Comptes Rendus Acad. Sci.*, Paris (1909), vol. 148, p. 1019.

BOYD, A. W. "Multiple Nest-building by Pied Wagtail," *British Birds*, September, 1928.

BUTTEL-REEPEN, H. Von. "Zur Psychobiologie der Hummeln," *Biol. Cent.* (1907), vol. 27, pp. 579 and 604.

BUYTENDIJK, F. J. J., and HAGE, J. "Sur la Valeur de Réaction de quelques Excitants Sensoriels simples dans la Formation d'une Habitude par les Chiens," *Arch. Néerl. Physiol.* (1923), vol. 8.

CORY, C. P. "The Habits of the Solitary Wasp (*Sceliphron Intrudens*)," *Journal of Bombay Natural History Society*, December, 1913.

CUNNINGHAM, BERT. *Axial Bifurcation in Serpents* (1937).

DE HAAN, J. A. BIERENS. *Animal Psychology for Biologists* (1929), "Probleme des tierischen Instinkts," *Naturwissenschaften* (1935).

DEMOLL, R. "Ueber die Vorstellungen der Tiere," *Zool. Jahrb., Zool. u. Physiol.* (1921), vol. 38, p. 405.

DEWAR, DOUGLAS. *Birds at the Nest* (1928).

DORSEY, GEORGE A. "A Red-headed Woodpecker Storing Acorns," *Bird-Lore* (1926), vol. 28, p. 333.

ENTEMAN, M. M. "On the Behaviour of Social Wasps," *Popular Science Monthly* (1902), vol. 61, p. 339.

FABRE, J. H. *The Wonders of Instinct* (1918).

FISHER, JAMES. *Birds as Animals* (1939).

FITZSIMONS, F. W. *Snakes* (1932).

GOOCH, G. B. "Instinct and Intelligence," *Contemporary Review,* August, 1939.

HATT, ROBERT T. "The Red Squirrel Farm," *Natural History,* May–June, 1929.

HAWKINS, JOHN L. "Multiple Nest-building by Blackbird," *British Birds,* August, 1925.

HINGSTON, R. W. G. *Problems of Instinct and Intelligence* (1928).

HOWARD, H. ELIOT. *The Nature of a Bird's World* (1935).

HUDSON, W. H. *Hampshire Days* (1903).

HUXLEY, JULIAN. *Essays of a Biologist* (1926); *Bird-watching and Bird Behaviour* (1930).

JOURDAIN, F. C. R. Editorial notes in *British Birds,* September, 1925, and July, 1937.

KATZ, DAVID. *Animals and Men* (1937).

KEARTON, RICHARD. *At Home with Wild Nature* (1922).

KIRKMAN, F. B. *Bird Behaviour* (1937); "The Gullibility of a Gull," *Country Life,* March 22, 1941.

LEVICK, G. MURRAY. *Antarctic Penguins* (1914).

MORGAN, C. L. *Animal Life and Intelligence* (1891).

NOBLE, RUTH CROSBY. *The Nature of the Beast* (1945).

PITT, FRANCES. *Animal Mind* (1927).

RITTER, WILLIAM E. *Animal and Human Conduct* (1928).

ROBINSON, E. KAY. *At Home with Nature* (1924).

RUSSELL, E. S. *The Behaviour of Animals* (1938).

SCHNEIRLA, T. C. "The Nature of Ant Learning," *Journal of Comparative Psychology* (1943), vol. 35, p. 149; "A Unique Case of Circular Milling in Ants," *American Museum Novitates,* No. 1253 (1944).

SETON, ERNEST THOMPSON. *Trail of an Artist-Naturalist* (1941).

SMITH, STUART. *How to Study Birds* (1945).

STURMAN-HUBLE, M., and STONE, C. P. "Maternal Behaviour in the Albino Rat," *Journal of Comparative Psychology* (1929), vol. 9, p. 234.

THOMSON, J. ARTHUR. *The Minds of Animals* (1927).

Tucker, B. W. "Brood-patches and the Physiology of Incubation," *British Birds*, July, 1943.

Washburn, Margaret F. *The Animal Mind* (1936).

Wheeler, W. M. *Ants, their Structure, Development and Behaviour* (1910).

Wiesner, Bertold P., and Sheard, Norah M. *Maternal Behaviour in the Rat* (1933).

Wolf, E. "The Homing Behaviour of Bees," *Journal of Social Psychology* (1930), vol. 1, p. 300.

7. Have Animals a Time-Sense?

Bannerman, David Armitage. *The Birds of Tropical West Africa* (1931).

Barrett, Charles. *Koonwarra* (1939).

Batten, H. Mortimer. *Nature from the Highways* (1925).

Beling, Ingeborg. "Ueber das Zeitgedächtnis der Biene," *Zeitschr. vergl. Physiologie* (1929), vol. 9, p. 259.

Bohn, G. "Sur les Movements Oscillatoires des *Convoluta roscoffensis*," *Compte Rendus Acad. Sc., Paris* (1903), vol. 137, p. 576; "Les *Convoluta roscoffensis* et la Théorie des Causes Actuelles," *Bulletin Museum d'Hist. Nat.* (1903), vol. 9, p. 352.

Bowles, J. Hooper. "Regularity in Habits of the Northwestern Flicker," *Murrelet* (1926), vol. 7, p. 42.

Burrows, William. "Periodic Spawning of 'Palolo' in Pacific Waters," *Nature*, January 13, 1945.

Buytendijk, F. J. J. *The Mind of the Dog* (1936).

Buytendijk, F. J. J., Fischel, W. and ter Laag, P. B. "Über den Zeitsinn der Tiere," *Arch. Néerl. Physiol.* (1935), vol. 20, p. 123.

Calhoun, John B. "Twenty-four Hour Periodicities in the Animal Kingdom," *Journal of Tennessee Academy of Science*, April and July, 1944; July and October, 1945; April, 1946.

Clark, Frances N. "The Life History of *Leuresthes tenuis*, an Atherine Fish with Tide-controlled Spawning Habits," *California Fish and Game Commission, Fish Bulletin* No. 10 (1925); "Grunion in Southern California," *California Fish and Game*, January, 1938.

Devany, J. L. "The Spruce Drummer," *Canadian Field-Naturalist*, January, 1921.

Eckstein, Gustav. *Everyday Miracle* (1940).

Forel, A. "La Mémoire du temps chez les Abeilles," *Bull. Institut Général Psychologique*, Paris (1906), vol. 6, p. 258.

Fox, H. Munro. *Selene or Sex and the Moon* (1928); *The Personality of Animals* (1940).

Fry, C. B. *Life Worth Living* (1939).

Funkhouser, Ray G. "Grunion Run To-night," *Nature Magazine,* June–July, 1940.

Gordon, Seton. *Hill Birds of Scotland* (1930).

Grabensberger, Wilhelm. "Untersuchungen über das Zeitgedächtnis der Ameisen und Termiten," *Zeitschr. vergl. Physiologie* (1933), vol. 20, p. 1; "Experimentelle Untersuchungen über das Zeitgedächtnis von Bienen und Wespen nach Verfütterung von Euchinin und Jodthyreoglobulin," *Zeitschr. vergl. Physiologie* (1934, *a.*), vol. 20, p. 338; "Der Einfluss von Salicylsäure, gelbem Phosphor und weissem Arsenik auf das Zeitgedächtnis der Ameisen," *Zeitschr. vergl. Physiologie* (1934, *b.*), vol. 20, p. 501.

Hertz, M. Quoted in *News Chronicle,* June 1, 1938.

Hudson, A. S. "Can Animals Count the Days?", *Popular Science Monthly,* November, 1888.

Hudson, W. H. *Adventures Among Birds* (1913).

Kalmus, Hans. "Ueber die Natur des Zeitgedächtnisses der Biene," *Zeitschr. vergl. Physiologie* (1934), vol. 20, p. 405; "Ueber das Problem der sogenannten Exogenen und Endogenen, sowie der erblichen Rhythmik und ueber organische Periodizitaet ueberhaupt," *Revista di Biologia* (1938), vol. 24, p. 191.

Katz, David. *Animals and Men* (1937).

Kleber, E. "Hat des Zeitgedächtnis der Bienen biologische Bedeutung?", *Zeitschr. vergl. Physiologie* (1935), vol. 22, p. 221.

Lockley, R. M. *I Know an Island* (1938).

Lutz, Frank E. *A Lot of Insects* (1941).

MacLeod, R. B., and Roff, M. F. "An Experiment in Temporal Disorientation," *Acta Psychologica* (1936).

Mann, William M. "Monkey Folk," *National Geographic Magazine,* May, 1938.

Park, Orlando. "Nocturnalism—the Development of a Problem," *Ecological Monographs* (1940), vol. 10, p. 485; "Concerning Community Symmetry," *Ecology* (1941), vol. 22, p. 164.

Pavlov, I. P. *Conditioned Reflexes* (1927).

Ratcliffe, Francis. *Flying Fox and Drifting Sand* (1938).

Ruch, F. L. "L'appréciation du temps chez le rat blanc," *Année Psychol.* (1931), vol. 32, p. 118.

Sanderson, Ivan. "In Tropic Tree-tops," *Animal and Zoo Magazine,* May, 1939.

SANDS, DOUGLAS B. "The Three-Hour Bird," *Bulletin of Massachusetts Audubon Society,* May, 1944.

SCHMID, BASTIAN. *Interviewing Animals* (1936).

THOMAS, SIR WILLIAM BEACH. Quoted in *Strand Magazine,* June, 1943; "The Open Air," *The Observer,* January 20, 1946.

THOMSON, BASIL. *The Indiscretions of Lady Asanath* (1898).

THOMSON, SIR J. ARTHUR. *The Minds of Animals* (1927).

VERRILL, A. HYATT. *Strange Birds and Their Stories* (1938).

VON FRISCH, KARL. "Methoden sinnephysiologischer und psychologischer Untersuchungen aus Bienen," *Handbuch biol. Arbeitsmethoden* (1921), vol. 5, p. 365; quoted in *Popular Science Monthly,* November, 1930.

VON STEIN-BELING, I. "Über das Zeitgedächtnis Bei Tieren," *Biological Reviews,* January, 1935.

WALKER, FRED. *Destination Unknown* (1934).

WASHBURN, MARGARET F. *The Animal Mind* (1936).

WOODWORTH, W. McM. "The Palolo Worm, *Eunice viridis* (Gray)," *Bulletin Museum of Comparative Zoology,* Harvard College, Cambridge, Mass., May, 1907.

8. THE MYSTERY OF BIRD ANTING

ADLERSPARRE, AXEL. *Ornith. Monatsber.,* September–October, 1936.

ALI, SĀLIM. "Do Birds Employ Ants to Rid Themselves of Ectoparasites?" *Journal of Bombay Natural History Society,* April, 1936.

AUDUBON, J. J. *Birds of America* (1844).

BATES, R. S. P. "Do Birds Employ Ants to Rid Themselves of Ectoparasites?" *Journal of Bombay Natural History Society,* April, 1937.

BUELL, BERTHA G. "A Blue Jay Apparently 'Anting,' " *Jack-Pine Warbler,* April, 1945.

CHISHOLM, A. H. *Bird Wonders of Australia* (1934); "Why Do Birds 'Ant' Themselves?" *Victorian Naturalist,* January, 1940; "The Problem of 'Anting,' " *Ibis,* July, 1944.

COLEMAN, EDITH. " 'Anting' by Birds," *Victorian Naturalist,* July, 1945.

DAVIS, MALCOLM. "A Robin Anting," *Auk,* April, 1944; "English Sparrow Anting," *Auk,* October, 1945.

FLOERICKE, K. *Ornith. Monatsber,* September–October, 1935.

FRAZAR, ABBOTT M. "Intelligence of a Crow," *Bulletin Nuttall Ornithological Club,* July, 1876.

FUNKE, DR. Quoted in *Ornith. Monatsber,* September–October, 1935.

GIVENS, T. V. "Bird 'Anting,'" *Victorian Naturalist,* May, 1945.

GROFF, MARY EMMA, and BRACKBILL, HERVEY. "Purple Grackles 'Anting' with Walnut Juice," *Auk,* April, 1946.

GROSKIN, HORACE. "Scarlet Tanagers 'Anting,'" *Auk,* January, 1943.

HAMPE, HELMUT. *Ornith. Monatsber,* September–October, 1935.

HEINROTH, O. *J. f. Ornith.* (1911), vol. 59, p. 172.

HILL, RAYMOND W. "Bronzed Grackle 'anting' with Mothballs," *Wilson Bulletin,* June, 1946.

HODGE, HERBERT. *It's Draughty in Front* (1938).

IVOR, H. R. "Further Studies of Anting by Birds," *Auk,* January, 1943; "Antics of Bird Anting," *Nature Magazine,* January, 1946.

LANE, FRANK W. "The Mystery of Bird 'Anting,'" *Country Life,* November 4, 1943.

LEWINGTON, P. G. Letter in *Country Life,* May 12, 1944.

McATEE, W. L. "'Anting' by Birds," *Auk,* January, 1938; "Red-eyed Towhee 'Anting,'" *Auk,* April, 1944.

NICE, MARGARET M., and TER PELKWYK, JOOST. "'Anting' by the Song Sparrow," *Auk,* October, 1940; "Cowbirds Anting," *Auk,* April, 1945.

NICHOLS, CHARLES K. "'Anting' by Robins," *Auk,* January, 1943.

OSMASTON, B. B. "Strange Behaviour of Certain Birds when in Possession of Strong Smelling Insects," *Journal of Bombay Natural History Society,* November, 1909; "Do Birds Employ Ants to Rid Themselves of Ectoparasites?" *Journal of Bombay Natural History Society,* December, 1936.

PILLAI, N. G. "Bird 'Bathing' in Ants," *Journal of Bombay Natural History Society,* December, 1941.

ROBIEN, PAUL. *Ornith. Monatsber,* September–October, 1935.

SHARP, DALLAS LORE. *Beyond the Pasture Bars* (1914).

STAEBLER, ARTHUR E. "A Robin Anting," *Wilson Bulletin,* September, 1942.

STEINIGER, FRITZ. "'Ekelgeschmack' und visuelle Anpassung, einige Futterungsversuche an Vögeln," *Zeitschr. f. Wiss. Zool.,* Ser. A (1937), vol. 149, p. 221.

STRESEMANN, E. "Werden Ameisen durch Vögel zum Vertreiben von Aussenparasiten benützt?", *Ornith. Monatsber,* July–August, 1935; "Die Benutzung von Ameisen zur Gefiederpflege," *Ornith. Monatsber,* September–October, 1935.

TEBBUTT, C. F. "'Anting' of Starling," *British Birds,* March, 1946.

THOMAS, RUTH HARRIS. "Catbird 'Anting' with a Leaf," *Wilson Bulletin*, June, 1946.

TROSCHÜTZ, ALFRED. "Bunte Bilder aus der Vogelstube," *Die gefiederte Welt*, October 8, 1931; *Ornith. Monastber*, September–October, 1935.

VAN TYNE, JOSSELYN. "'Anting' by the Robin and Towhee," *Auk*, January, 1943.

WHEELER, WILLIAM MORTON. *Ants, Their Structure, Development and Behaviour* (1910).

WRIGHT, HORACE W. "A Nesting of the Blue-Winged Warbler in Massachusetts," *Auk*, October, 1909.

ZIMMER, JOHN T. Quoted in *Natural History*, April, 1946.

9. THE LEGEND OF THE HEDGEHOG AND THE FRUIT

B. L. "An Original Mode of Gathering Fruit," *Die Gartenlaube* (1865), quoted in *Science Gossip*, August, 1867.

BRIGGS, J. J. "A Fauna of Melbourne," *Zoologist* (1848), vol. 6, p. 2278.

CHRISTY, MILLER. "The Ancient Legend as to the Hedgehog Carrying Fruits Upon its Spines," *Memoirs and Proceedings of Manchester Literary and Philosophical Society*, May, 1919; February, 1923.

COOKE, MORDECAI C. "Hedgehog Eccentricities," *Science Gossip*, October, 1867.

DARWIN, CHARLES, "Hedgehogs," *Science Gossip*, December, 1867.

DOVETON, F. B. "Hedgehog," *Nature Notes*, June, 1904.

DRUCE, GEORGE C. "Mediæval Bestiaries," *Journal of British Archæological Association* (1920), vol. 26, p. 35.

GUDGER, E. W. "Does the Jaguar Use His Tail as a Lure in Fishing?" *Journal of Mammalogy*, February, 1946.

HEATHCOTE, B. T. Letter in *The Field*, February 3, 1945.

KING OF THE NORFOLK POACHERS. *I Walked by Night* (1935).

KNIGHT, M. FORSTER. "Hedgie Keats; A Tame Hedgehog," *Country Life*, February 6, 1942.

LANE, FRANK W. "The Hedgehog and Fruit Legend," *The Field*, December 12, 1944.

LIU, CH'ENG-CHAO. "Notes on the Food of Chinese Hedgehogs," *Journal of Mammalogy*, August, 1937.

MOLL, F. "Curious Habit of the Hedgehog," *Australian Mus. Mag.*, October–December, 1924.

PARKER, ERIC. *Oddities of Natural History* (1943).

PEACOCK, MABEL. "Hedgehogs Carrying Fruit," *Science Gossip,* August, 1897.

PITT, FRANCES. "The Hedgehog," in *The Romance of Nature* (1937).

SPICER, W. W. "Hedgehogs," *Science Gossip,* March, 1868.

WARNER, W. H. "Hedgehog," *Nature Notes,* July, 1904.

10. BIRD HITCH HIKERS

ALEXANDER, W. B. "The Woodcock in the British Isles," *Ibis,* January, 1946.

BAILEY, FLORENCE MERRIAM. "The Scissor-tailed Flycatcher in Texas," *Condor,* March–April, 1902.

BAKER, SIR SAMUEL W. *The Nile Tributaries of Abyssinia* (1867).

BAKER, E. C. STUART. "The Game Birds of India, Burma and Ceylon," *Journal of Bombay Natural History Society,* December, 1923.

BENT, ARTHUR CLEVELAND. *Life Histories of North American Wildfowl* (1923 and 1925); *Life Histories of North American Shore Birds* (1927); *Life Histories of North American Gallinaceous Birds* (1932); *Life Histories of North American Birds of Prey* (1937 and 1938). (U.S. National Museum.)

BOULENGER, E. G. "Some Animals Take Rides," *John o' London's Weekly,* January 6, 1939.

BRANDT, HERBERT. *Texas Bird Adventures* (1940).

CLAYPOLE, E. W. "Migration of the Wagtail," *Nature,* February 24, 1881 (quoting *The Evening Post,* New York).

CUMING, E. D. *Idlings in Arcadia* (1936).

DYER, THISELTON. *English Folk-Lore* (1884).

GARRETSON, MARTIN S. *The American Bison* (1938).

GLEGG, WILLIAM E. "Animate Perching Associations," *Ibis,* October, 1945.

GUILLEMARD, F. H. H. *The Cruise of the 'Marchesa'* (1889).

HAIG-THOMAS, DAVID. *Time and Tide,* December 9, 1939.

HARTING, JAMES E. *The Recreations of a Naturalist* (1906).

INGERSOLL, ERNEST. *The Wit of the Wild* (1906).

LANE, FRANK W. "Do Birds Fly Pick-a-back?" *The Field,* September 15, 1945.

LINCOLN, FREDERICK C. "Notes on the Bird Life of North Dakota," *Auk,* January, 1925.

LLOYD, G. W. Letter in *The Field,* October 20, 1945.

LOWERY, GEORGE H. junior. "Evidence of Trans-Gulf Migration," *Auk,* April, 1946.

MACKAY, GEORGE H. "Behaviour of a Sandhill Crane," *Auk,* July, 1893.

McATEE, W. L. "Birds Pick-a-back," *Scientific Monthly*, March, 1944.

MERRILL, J. C. "The Crane's Back," *Forest and Stream*, March 10, 1881.

MYERS, J. G. *Dr. J. G. Myers on Migrants and on Bird-insect Nesting Associations in the Sudan* (communicated by R. E. Moreau), *Ibis*, January, 1943.

NELSON, T. H. *The Zoologist*, February, 1882.

NEUMANN, ARTHUR H. *Elephant-Hunting in East Equatorial Africa* (1898).

NORTH, M. E. W. "The Use of Animate Perches by the Carmine Bee-eater and Other African Species," *Ibis*, April, 1944. Further notes on the subject appeared in subsequent issues.

OLIVER, JAMES A. "An Aggregation of Pacific Sea Turtles," *Copeia*, July, 1946.

RAE, JOHN. Quoted in *Canadian Record of Science* (1888), vol. 3, p. 126.

SERVICE, ROBERT. "The Waders of Solway," *Trans. Natural History Society of Glasgow* (1908), vol. 8, p. 48.

SETON, ERNEST THOMPSON. *Lives of Game Animals* (1927).

THOMAS, SIR WILLIAM BEACH. *The Yeoman's England* (1934).

VAN-LENNEP, HENRY J. *Bible Lands* (1875).

VERRILL, A. HYATT. *Strange Birds and Their Stories* (1938).

WIGHTWICK, DUDLEY. Letter in *John o' London's Weekly*, November 26, 1937.

WILLIAMSON, HENRY. *The Story of a Norfolk Farm* (1941).

WILLIAMSON, W. E. Letter in *The Field*, January 19, 1946.

11. SOME EXPERIMENTS WITH ANIMALS

ALLAN, P. B. M. *Talking of Moths* (1943).

ARUNDEL, REGINALD. "Police Dogs I Have Known," *Chambers's Journal*, July, 1945.

BARTON, D. R. "Attorney for the Insects" (Frank E. Lutz), *Natural History*, October, 1941.

BAYER, E. "Beiträge zur Zweikomponontentheorie des Hungers," *Zeitschr. f. Psychologie* (1929), vol. 112.

BETHE, A. "Dürfen wir den Ameisen u. Bienen psychische Qualitäten zuschreiben?", *Pflügers Arch.* (1898), vol. 70, p. 15; "Noch einmal über d. psychischen Qualitäten der Ameisen," *Pflügers Arch.* (1900), vol. 79, p. 39.

BURR, MALCOLM. *The Insect Legion* (1939).

BUYTENDIJK, F. J. J. *The Mind of the Dog* (1936).

CARR, H. A., and WATSON, J. B. "Orientation in the White Rats," *Journal of Comparative Neurology and Psychology* (1908), vol. 18, p. 27.

CRILE, GEORGE. *Intelligence, Power and Personality* (1941).

ELTRINGHAM, H. *The Senses of Insects* (1933).

ENGELMANN, W. "Untersuchungen über Schallokalisation bei Tieren," *Zeitschr. f. Psychologie* (1928), vol. 105, p. 317.

FOX, H. MUNRO. *The Personality of Animals* (1940).

FRANÇON, JULIEN. *The Mind of the Bees* (1939).

GLICK, P. A. "The Distribution of Insects, Spiders, and Mites in the Air," *Technical Bulletin* No. 673 (1939), U.S. Dept. of Agriculture.

HARLOW, HARRY F. and BROMER, JOHN A. "Comparative Behaviour of Primates," *Journal of Comparative Psychology*, October, 1939.

HECK, L. "Ueber die Bildung einer Assoziation beim Regenwurm auf Grund von Dressurversuchen," *Med. naturwiss. Zeit. Lotos* (1920), vol. 68, p. 168.

KATZ, DAVID. *Animals and Men* (1937).

KELLER, H., and BRÜCKNER, G. H. "Neue Versuche über das Richtungshören des Hundes," *Zeitschr. f. Psychologie* (1932), vol. 126, p. 14.

KELLOGG, VERNON L. "Some Silkworm Moth Reflexes," *Biological Bulletin*, February, 1907.

KETTLEWELL, H. B. D. "Female Assembling Scents with Reference to an Important Paper on the Subject," *Entomologist*, January, 1946.

LECKY, PRESCOTT. "Even Worms Solve Puzzles," *Popular Science Monthly*, August, 1927.

LOESER, JOHANN A. *Animal Behaviour* (1940).

LUBBOCK, SIR JOHN. *On the Senses, Instincts and Intelligence of Animals* (1888).

LUTZ, FRANK E. "Experiments with 'Wonder Creatures'," *Natural History*, March–April, 1929; *A Lot of Insects* (1941).

MAIER, N. R. F., and SCHNEIRLA, T. C. *Principles of Animal Psychology* (1935).

MARAIS, EUGENE N. *The Soul of the White Ant* (1937).

MÖBIUS, K. "Die Bewegungen der Thiere und ihr psychischer Horizont," *Schrift d. Naturwiss. vergl. Schleswig-Holstein* (1873), vol. 1, p. 113.

MOORE, A. U., and MARCUSE, F. L. "Salivary, Cardiac and Motor Indices of Conditioning in two Sows," *Journal of Comparative Psychology*, February, 1945.

MOWRER, O. H. "Animal Studies in the Genesis of Personality,"

Trans. New York Acad. Science, Ser. 2, vol. 3, No. 1 (1940); quoted in *The Reader's Digest,* November, 1939.

MURCHISON, CARL. Editor; *A Handbook of General Experimental Psychology* (1934).

NOBLE, G. K. "Probing Life's Mysteries," *Natural History,* September–October, 1930; "Courtship and Sexual Selection of the Flicker (*Colaptes auratus luteus*)," *Auk,* July, 1936.

NOBLE, RUTH CROSBY. *The Nature of the Beast* (1945).

OPFINGER, E. "Über die Orientierung der Biene an der Futterquelle," *Zeitschr. vergl. Physiologie* (1931), vol. 15, p. 431.

RILEY, C. V. "The Senses of Insects," *Nature,* June 27, 1895.

ROBERTS, AUSTIN. *The Birds of South Africa* (1940).

ROMANES, G. J. *Animal Intelligence* (1883); "Experiments on the Sense of Smell in Dogs," *Nature,* July 21, 1887.

RUSSELL, E. S. *The Behaviour of Animals* (1938).

SCHMID, BASTIAN. *Interviewing Animals* (1936).

SCHNEIRLA, T. C. "The Nature of Ant Learning," *Journal of Comparative Psychology,* April, 1943.

SHASTID, THOMAS H. *Our Own and Our Cousins' Eyes* (1926).

SMITH, STUART. *How to Study Birds* (1945).

TEALE, EDWIN WAY. *The Golden Throng* (1942).

THORPE, W. H. "A Type of Insight Learning in Birds," *British Birds,* July, 1943.

TORREY, HARRY B. "Animal Experimentation," *Scientific Monthly,* August, 1939.

TRIPLETT, N. B. "The Educability of the Perch," *American Journal of Psychology* (1901), vol. 12, p. 354.

VERLAINE, L. "L'instinct et l'intelligence chez les oiseaux," *Recherches Philosophie* (1934), vol. 3, p. 285.

VON FRISCH, KARL. "Ueber den Geruchsinn der Biene," *Zool. Jahrb., Zool. u. Physiol.* (1919), vol. 37, p. 1; "Über die Sprache der Bienen," *Zool. Jahrb., Zool u. Physiol.* (1923), vol. 20, p. 1; "The Language of Bees," *Science Progress,* July, 1937; *Bees* (1951).

WALLS, GORDON LYNN. *The Vertebrate Eye* (1942).

WASHBURN, MARGARET F. *The Animal Mind* (1936).

YERKES, R. M. "The Intelligence of Earthworms," *Journal of Animal Behaviour* (1912), vol. 2, p. 332.

12. HOW THE PASSENGER PIGEON BECAME EXTINCT

ABBOTT, ROY L. "The Passenger Pigeon," *Natural History,* February, 1944.

ALLEE, W. C. *The Social Life of Animals* (1938).

ALLEN, ARTHUR A. In Preface to James T. Tanner's *The Ivory-Billed Woodpecker* (1942).

ANBUREY, THOMAS. *Travels through the Interior Parts of America* (1789).

AUDUBON, JOHN JAMES. *The Bird of America* (1844).

BREWSTER, WILLIAM. "The Present Status of the Wild Pigeon (*Ectopistes migratorius*) as a Bird of the United States, with some Notes on its Habits," *Auk*, April, 1889.

BUCKINGHAM, J. S. *The Slave States of America* (1842).

CHILDS, JOHN LEWIS. "Personal Recollections of the Passenger Pigeon," *The Warbler*, July, 1905.

DUNLOP, WILLIAM ("A Backwoodsman"). *Statistical Sketches of Upper Canada for the Use of Emigrants* (1832).

FEATHERSTONHAUGH, G. W. *Excursion through the Slave States* (1844).

FENTON, WILLIAM N., and DEARDORFF, MERLE H. "The Last Passenger Pigeon Hunts of the Cornplanter Senecas," *Journal Washington Academy of Science*, October, 1943.

FISHER, JAMES. *Watching Birds* (1940).

FLEMING, C. A. "Passenger Pigeons," *Owen Sound Daily Sun Times*, April 11, 1931.

FORBUSH, EDWARD HOWE. *Birds of Massachusetts and Other New England States* (1927).

GLADSTONE, HUGH S. *Record Bags and Shooting Records* (1930).

HODGE, C. F. "A Last Word on the Passenger Pigeon," *Auk*, April, 1912.

KALM, PEHR. "A Description of the Wild Pigeons which Visit the Southern English Colonies in North America, During Certain Years, in Incredible Multitudes," *Auk*, January, 1911 (translated by S. M. Gronberger from a Swedish article published in Stockholm in 1759).

KING, W. Ross. *The Sportsman and Naturalist in Canada* (1866).

LOWE, P. R. "A Reminiscence of the Last Great Flight of the Passenger Pigeon (*Ectopistes migratorius*) in Canada" (from information supplied by Dr. A. B. Welford), *Ibis*, January, 1922.

MARTIN, E. T. "Among the Pigeons—A Reply to Professor Roney's Account of the Michigan Nesting of 1878," *American Field*, January 25, 1879.

MATHER, COTTON. Quoted in *Auk*, October 1944; April, 1945.

MEASE, JAMES. *A Geological Account of the United States* (1807).

MERSHON, W. B. *The Passenger Pigeon* (1907).

MITCHELL, MARGARET H. "The Passenger Pigeon in Ontario" (1935), *Contribution No. 7 of the Royal Ontario Museum of Zoology.*
PENNANT, THOMAS. *Arctic Zoology* (1785).
POKAGON, SIMON. "The Wild Pigeon in North America," *The Chautauquan,* November, 1895.
RONEY, H. B. "A Description of the Michigan Pigeon Nesting of 1878," *Chicago Field,* January 11, 1879.
THWAITES, REUBEN GOLD (translator). *Jesuit Relations and Allied Documents* (for 1662–1663) (1901).
TOWNSEND, CHARLES WENDELL. "Passenger Pigeon," in *U.S. National Museum Bulletin* No. 162 (1932).
WILSON, ALEXANDER. *American Ornithology* (1832).
WILSON, ETTA S. "Personal Recollections of the Passenger Pigeon," *Auk,* April, 1934.
WOOD, WILLIAM. *New England Prospect* (1635).
WRIGHT, ALBERT HAZEN. "Early Records of the Passenger Pigeon," *Auk,* October, 1910; July, 1911; October, 1911.

13. How THE BUFFALO WAS SAVED

ALLEN, J. A. "The American Bisons, Living and Extinct," *Memoirs of the Museum of Comparative Zoology,* Cambridge, Mass. (1876), vol. 4, No. 10.
BARCLAY, EDGAR N. *Big Game Shooting Records* (1932).
BRANCH, E. DOUGLAS. *The Hunting of the Buffalo* (1929).
BROWN, BRUCE. "Railway Excursion and Buffalo Hunt," *Natural History,* May, 1942.
CAHALANE, VICTOR H. "Restoration of Wild Bison," *Transactions of the Ninth North American Wildlife Conference* (1944); "Buffalo Go Wild," *Natural History,* April, 1944; "Buffalo Wild or Tame?" *American Forests,* October, 1944.
CASTAÑEDA, PEDRO DE. *The Journey of Coronado,* 1540–1542, Winship translation (1922).
CATLIN, GEORGE. *North American Indians* (1913).
CODY, WILLIAM F. *Life of Buffalo Bill, by Himself* (1879).
DODGE, RICHARD IRVING. *The Plains of the Great West and Their Inhabitants* (1877).
ELLSWORTH, LINCOLN. "The Last Wild-Buffalo Hunt," in *Explorer's Club Tales* (1937).
GARRETSON, MARTIN S. *The American Bison* (1938).
GREY, ZANE. *The Thundering Herd* (1925).
HENRY, ALEXANDER. *Journal in Red River Valley,* 1799–1814 (1897).

HEWITT, C. GORDON. "The Coming Back of the Bison," *Natural History*, December, 1919.

HODGE, F. W., and LEWIS, T. H. *Spanish Explorers in the Southern United States*, 1528–1543 (1907).

HORNADAY, WILLIAM T. "The Extermination of the American Bison," *Annual Report of the Smithsonian Institution* for 1887 (1889), Part 2.

HULBERT, A. B. *Historic Highways of America* (1902).

INMAN, HENRY. *Tales of the Trail* (1917).

IRVING, WASHINGTON. *The Adventures of Captain Bonneville, U.S.A., in the Rocky Mountains* (1895).

LANE, FRANK W. *The Elements Rage* (1945).

LEWIS, MERIWETHER, and CLARK, WILLIAM. *Original Journals of the Lewis and Clark Expedition*, 1804–1806 (1904–1905).

McCRACKEN, HAROLD. "The Sacred White Buffalo," *Natural History*, September, 1946.

MEAD, JAMES R. "Some Natural-History Notes of 1859," *Trans. Kansas Academy of Science*, June, 1899.

MOORE, ELY. "A Buffalo Hunt with the Miamis in 1854," *Kansas Historical Collections* (1908), vol. 10, p. 402.

PEATTIE, DONALD CULROSS. "America's Greatest Host," *Natural History*, October, 1943.

ROLLINS, P. A. *The Cowboy* (1922).

ROSS, ALEXANDER. *The Fur Hunters of the Far West* (1855); *Red River Settlement* (1856); *Alexander Ross's Adventures of the First Settlers on the Oregon or Columbia River*, 1810–1813 (1904).

SETON, ERNEST THOMPSON. *Life Histories of Northern Animals* (1909); *Lives of Game Animals* (1927).

SMITH, HELENA HUNTINGTON. "Want a Buffalo?" *Collier's Magazine*, August 12, 1944.

TOWNSEND, JOHN K. *Narrative of a Journey across the Rocky Mountains to the Columbia River*, 1833–1834 (1905).

WILLIAMSON, GLENN YERK. "Thunder in the West," *Fauna*, March, 1946.

YOUNG, STANLEY P., and GOLDMAN, EDWARD A. *The Wolves of North America* (1944).

YOUNG, STANLEY P. *Sketches of American Wildlife* (1946).

INDEX

Index

Pig, 122
Pig-tailed macaque, 120
Pigeon, 57
Pigeon, homing, 198
Pigeon passenger, 66, 71–72, 236–254
Pike, 45–46, 67, 85, 225
Pileated tinamou, 149
Pin-tail sandgrouse, 149
Pine-warbler, 74
Pirarucu, 105
Plover, 30
Plover, Egyptian, 191
Pony, forest, 92
Porcupine, 121
Porcupine fish, 104–105
Processionary caterpillar, 127
Psammophis, 139
Purple grackle, 170
Puffin, 69, 102

Q

Quail, 29, 70
Quail, grey, 198

R

Rabbit, 82, 91–92, 103
Rail, California clapper, 87
Rat, 89, 92, 137, 157–158, 182, 203, 226, 230–231
Rattlesnake, 25–26, 89–90
Raven, 57
Ray, 105
Red-browed finch, 171
Red-tailed hawk, 214
Reef heron, Australian, 147–148
Rhinoceros, 191
Rhynchota, 170
Rice bird, 85
Robin, 72, 84, 101, 165–166, 175–176
Robin, Asiatic Pekin-, 175
Rock-bee, 90
Rook, 63, 215
Rose-breasted grosbeak, 169
Ruby-throated humming-bird, 84, 201
Ruffed grouse, 71

S

Sabre-toothed tiger, 92
Sailfish, 48
Salamander, 89
Salmon, 47, 242
Sand-snake, 139
Sandgrouse, pin-tail, 149
Sandhill crane, 196
Sandpiper, 86
Saw-belly, 67
Scarites pyracmon, 110
Scissor-tailed flycatcher, 193–194
Scops owl, 73
Sea-anemone, 151
Sea-urchin, 121
Seagull, 49, 52–53
Seal, 33–34
Setter, dog, 207–208
Shark, 97, 105–106, 154
Shark, whale, 93
Sharp-shinned hawk, 30
Sheep, 56, 75, 92, 138–139, 142, 192
Short-eared owl, 201–204
Silkworm, 122
Silkworm, Japanese ailanthus, 209
Silverside, 154
Six-lined race-runner lizard, 29
Skua, 201
Skunk, 122
Skylark, 95
Sloth, giant, 92
Snail, 83
Snake, 31, 89, 92
Snake, sand, 139
Snipe, 29, 86
Song sparrow, 170
Sooty tern, 55
South African yellow weaver-bird, 229
Sow, 122
Sparrow, 50, 85, 102, 146, 200
Sparrow, song, 170
Sparrowhawk, 20, 30–31
Spider, 33, 84–85, 88–90, 111–114, 231
Spider, Argentine, 89
Spider, Australasian, 111–114

Y

Z

NOTES

NOTES

NOTES

NOTES

NOTES

NOTES